Malik Osmane

Rift-lines within european regulatory framework for biosimilars when taking heterogeneity and variation during lifecycle of the reference biologic and the biosimilar into account

Anchor Academic
Publishing

Osmane, Malik: Rift-lines within european regulatory framework for biosimilars when taking heterogeneity and variation during lifecycle of the reference biologic and the biosimilar into account. Hamburg, Anchor Academic Publishing 2013

Buch-ISBN: 978-3-95489-187-0
PDF-eBook-ISBN: 978-3-95489-687-5
Druck/Herstellung: Anchor Academic Publishing, Hamburg, 2013

Bibliografische Information der Deutschen Nationalbibliothek:
Die Deutsche Nationalbibliothek verzeichnet diese Publikation in der Deutschen Nationalbibliografie; detaillierte bibliografische Daten sind im Internet über http://dnb.d-nb.de abrufbar.

Bibliographical Information of the German National Library:
The German National Library lists this publication in the German National Bibliography. Detailed bibliographic data can be found at: http://dnb.d-nb.de

Hermannstal 119k, 22119 Hamburg
http://www.diplomica-verlag.de, Hamburg 2013
Printed in Germany

Abstract

Introduction:

Biopharmaceutical medicinal products (biologics) represent a huge financial market. Thus upon patent protection expiry of the innovator (reference) biologic there is interest from industry to gain a portion of this market by launching a "similar" biologic at a reduced development cost, thus boosting potential gains.

The EMA responded to this desire and lead the guidance process with industry on the topic of biosimilars. This guidance process can be described as a responsive or accompanying process, which resulted in the successive development of three (hierarchical) layers of regulatory guidance documentation for biosimilars by the EMA. The lowest layer (product class specific) was until recently still being updated.

Based on the experience gained with biosimilars in the past, the EMA started to introduce a second generation series of guidance documents, which take into account the past, current and possibly future challenges of biosimilars. Those proposals were evaluated by EMA and partially incorporated into new guidance documents.

This work highlights the challenges and risks associated with biosimilar submissions for large and complex bio-molecules such antibodies. The following topics were analysed, related to the overall biosimilar concept and its implicated costs:

1) ***What are the implications of heterogeneity and variation through the life cycle of the biosimilar and the reference biologic, from a European perspective?*** - refer to chapter 4.
2) ***What should be the scope of trials for biosimilars?*** - refer to chapter 5.
3) ***Why is extrapolation of indications for biosimilar controversial?*** - refer to chapter 6.

Results:

1) As the tile of this study indicates, there are unaddressed questions for the regulator with regard to the unsolved dynamic of heterogeneity and variations of the quality profile, which have potential implications on safety and efficacy. This is neglected and not taken into account seriously enough by the stakeholders.

2) Non-inferiority trials just barely meet the biosimilarity requirements (in my view), and equivalence studies require that the quality profiles of the biosimilar falls within the quality profile of the reference biologic, to have enough confidence to accomplish successfully the clinical trial phase of the comparability exercise.

However, regardless of the trial design, the current approach that heterogeneity and variation of the reference biologic is controlled by the oversight of the EMA, and therefore as an extension changes in the quality profile of the reference biologic will result in the same safety and efficacy pattern as the pre-changed is lacking foresight of potential problems during the biosimilar development.

Thus during the course of this study proposals and clarifications were submitted to the EMA and partially adopted by the EMA.

3) A biosimilar that may have been matched to the pre-change quality profile of the reference biologic may struggle to show similarity in its clinical profile when compared to the post-change quality profile of the reference biologic. It is unreasonable to assume per se similar safety and

efficacy from different quality profiles; thus, extrapolation of indication in those circumstances should not be allowed.

Solution:

Further, the only (in my view) progressive way to deal with such foreseeable situations from the biosimilar developer's point of view is to incorporate a design space. To that, in fact to any "sudden" change of the quality profile of the reference biologic, the biosimilar developer can respond in due time, assuring that the material used for the clinical trial comparability exercise is in fact within the boundaries of the quality profile of the reference biologic.

For particular scenarios where the heterogeneity and variation of the reference biologic, even without a major change, results in a detectable difference in the clinical phase, two possible causes need to be investigated.

1. The quality profile of the reference biologic and the biosimilar are not similar enough.
2. Trials would also fail, when different batches of the biosimilar would be compared to itself.

The reasons for such failure could be manifold, such as selection of the margin of tolerance and related to the selection of the clinical endpoint, or that the reference biologic is in its performance not consistent, i.e. the efficacy margins are low compared to the placebo arm, for example.

For these anticipated scenarios more guidance is needed, which will possibly only be available once the EMA is confronted with such a situation. One can only hope that that information will be made public for further analysis.

Table of Contents

Note:
The research chapter 4, chapter 5 and chapter 6 deal individually with dedicated topics. Thus for each of those chapters an individual introduction, method and material parts, results and discussion section was written. Based on that approach information obtained from questionnaires or feedback received from submissions to EMA was split based on its content into those research chapter 4, chapter 5 and chapter 6.
The chapters general aspects and literature research was carried out as part of chapter 1, chapter 2 and chapter 3.
An integrated discussions and conclusions were performed as part of chapter 7 and chapter 8.

For Yasmina

and Nouara

to my beloved daughters

Acknowledgements

I would like to cite the following people for their help in the review process of this study:

Joe Brady, PhD:

For the endless help guidance and motivation with his exceptionally pleasant and routined way of dealing with people.

For helping me to challenging my logic and train of thoughts.

Regina Martins:

For helping me to challenging my logic and train of thoughts.

Jolyon Dodgson, PhD:

For helping me to challenging my logic and train of thoughts.

For helping me formatting parts of the study.

Paul Declerck, PhD Professor:

For helping me to challenging my logic and train of thoughts.

All other experts who supported my work through providing literature free of charge, for commenting on and reviewing my questionnaires.

This study was carried out with no other means and input beside the ones cited in this study. M. Osmane is the genuine author of this study including all its concepts and ideas.

(In the case where those ideas were already voice by others, this did not happen intentionally and was not discovered as part of my literature review.)

List of abbreviations

Abbreviation	Definition
AARP	The American Association of Retired People
Ab	Antibody
ADCC	Antibody-dependent Cellular Cytotoxicity
BMWP	Biological Monitoring Working Party
BPCIA	Biologics Price Competition and Innovation Act
BWP	Biologics Working Party
CHMP	Committee for Human Medicinal Products
CHO	Chinese Hamster Ovary
CO_2	Carbon Dioxide
DIT	Dublin Institute of Technology
DNA	Deoxyribonucleic Acid
EEA	European Economic Area
EFTA	European Free Trade Association
EMA	European Medicines Agency
EMEA	European Medicines Agency (former name)
EPAR	European Public Assessment Report
ER	Endoplasmatic Reticulum
EU	European Union
FDA	Food and Drug Administration
HPLC	High-performance Liquid Chromatography
ICH	International Conference on Harmonization
IgG	Immunoglobulin G
kDA	Kilodalton
MA	Marketing Authorization
MAA	Marketing Authorization Application
mAbs	Monoclonal Antibodies
MHLW	Ministry of Health, Labour and Welfare
mRNA	Messenger Ribonucleic Acid
NCE	New Chemical Entity
O_2	Oxygen
PCT	Patent Cooperation Treaty
PD	Pharmacodynamic
PK	Pharmacokinetic
PRCA	Pure Red Cell Aplasia
QA	Quality Assurance
QA	Quality Attributes
QTPP	Quality Target Product Profile
RNA	Ribonucleic Acid

RPB	Reference Biotherapeutic Product
SPC	Summary of Product Characteristics
TOPRA	The Organization for Professionals in Regulatory Affairs
US	United States
UV	Ultra Violet
VEGF	Vascular Endothelial Growth Factor
WHO	World Health Organization

Chapter 1: Introduction

1.0. Rationale on the selection of the topic (as guide for future students)

I started this course in pharmaceutical QA being aware that I would not benefit from full sponsorship from my employer. Although my financial possibilities were limited at the time, I wanted to explore the opportunity to take a step forward in my career in a meaningful way, by improving and developing my knowledge. Now, when I am close to finishing my study, looking back I realized it was the right choice as it was the gateway to a better future not only from the point of view of my knowledge but also from a personal perspective.

My educational and scientific background is in immunology[1], as I graduated with a Diploma thesis at the Max Planck Institute for Immunology in Freiburg. I wanted to use this scientific background to find a proper topic for my study project. On the other hand, I also wanted to focus on a topic I am personally interested in and that will help me find better career opportunities.

The search for a proper study topic started very early. I was asked by my teachers where do I want to be in 5-years time and what topic really interested me, and yet would allow me to gain further competencies and scientific skills. Finding the answer to this question took some time, but going through this process was deeply revealing and gave me clarity about my future career path.

At the end of 2010, following a lecture and an article in TOPRA, I found what I was looking for: Biosimilars.

The Biosimilar topic represents a good symbiosis of the DIT course content, focused on EMA regulations, and in accordance with my background in immunology. It is an emerging top regulatory topic, as the generic biologics industry (the correct term is biosimilar industry) seeks to gain access to the highly profitable biologics[2] market.

There was a good likelihood of writing a meaningful research paper and conducting an interesting study. The regulatory framework within Europe (the leading regulatory region for biosimilars) continues to require clarification and is in need of further development, especially at the theoretical level.

---

[1] Immunology:
The branch of medicine and biology concerned with immunity. Oxford Online Dictionary.

[2] Biologic:
A) Relating to biology; biological: there is growing interest in the biologic activities of plant extracts in the treatment of disease; or
B) Another term for biological (noun): these natural biologics can be as potent as manufactured drugs. Oxford Online Dictionary.

By commenting on EMA regulations and directly interacting with the EMA, this study bears the potential to bring a new perspective to the European biosimilar regulations, overall an excellent strategic opportunity for the elaboration of this study.

1.1 Definitions

To understand the notion of biosimilar and differences to classical small molecules, it is first required to outline the term generic.

The term 'generic' medicinal product is used to describe chemicals, which are small molecules. They are structurally equivalent to an innovator medicinal product whose patent and/or data protection period has expired.

Bioequivalence of the generic medicine with a reference product is required, which usually does not include clinical trial, or in the case where it does they are minimal.

It is permissible for generic companies to apply for a M.A., due to the pharmacopeia data from the reference medical product outline specification and the ability to produce robustly the same molecular entity, through synthesis. This high chemical purity product is very much different to biotherapeutics/biologics which are relatively large and complex proteins that are difficult to characterize (1).

1.1.1 Definition of biological medicine[3] and major differences to "classical" chemical medicine

Biologics refers to naturally occurring substances generated by animals or microorganisms, or that can be created through biotechnology[4] — recombinant DNA technology[5].

Biological medicines are medicinal products, which are produced using a living system or organism. Typically modern biotechnology products are proteins, since almost all life on earth is based on proteins with its manifold functional groups, thus being an ideal basis for drug development. Proteins are produced in cells using recombinant DNA technology. From the

---

[3] Biological medicine:
A medicine that contains one or more active substances made by or derived from a biological source. Questions and answers on biosimilar medicines, EMA.

[4] Biotechnology:
The exploitation of biological processes for industrial and other purposes, especially the genetic manipulation of microorganisms for the production of antibiotics, hormones, etc.. Oxford Online Dictionary.

[5] Recombinant DNA technology:
Joining together of DNA molecules from two different species that are inserted into a host organism to produce new genetic combinations that are of value to science, medicine, agriculture and industry. Britannica Academic Edition.

European regulatory perspective, they are defined as *medicinal products*[6] that are developed based on biotechnological processes:

- Recombinant DNA technology
- Controlled expression of genes[7] coding for biologically active proteins in prokaryotes[8] and eukaryotes[9] including transformed mammalian cells
- Hybridoma[10] and monoclonal antibody[11] methods

Biological medicines differ from "classical" chemical medicines fundamentally through the following perspective:

1) Chemical medicines are usually small molecules, which are produced using a chemical synthesis process. Aspirin, for example, has a molecular weight of ca. 0.2 kDA. Biological medicines are usually much larger. Interferon alpha, for example, has a molecular size of ca. 19kDA and the IgG molecule has a molecular size of ca. 150kDA.

2) This direct molecular size/weight difference between biological medicines (later abbreviated as biologics) and chemical medicines, results also in a more complex three-dimensional structure of biologics.

3) The complex structures of biologics are more prone to vary compared to chemical medicines and poses inherent heterogeneity[12], refer to section: 2.2 Life cycle in relation to heterogeneity and variation for details on that heterogeneity and variation issue.

---

[6] Medicinal product:

(a) Any substance or combination of substances presented as having properties for treating or preventing disease in human beings; or

(b) Any substance or combination of substances which may be used in or administered to human beings either with a view to restoring, correcting or modifying physiological functions by exerting a pharmacological, immunological or metabolic action, or to making a medical diagnosis.Article 1, DIRECTIVE 2001/83/EC

[7] Gene:

A distinct sequence of nucleotides forming part of a chromosome, the order of which determines the order of monomers in a polypeptide or nucleic acid molecule which a cell (or virus) may synthesize. Oxford Online Dictionary.

[8] Prokaryote:

A microscopic single-celled organism which has neither a distinct nucleus with a membrane nor other specialized organelles, including the bacteria and cyanobacteria. Oxford Online Dictionary.

[9] Eukaryote:

An organism consisting of a cell or cells in which the genetic material is DNA in the form of chromosomes contained within a distinct nucleus. Eukaryotes include all living organisms other than the eubacteria and archaea. Oxford Online Dictionary.

[10] Hybridoma:

The fusion of a myeloma cell from a line that has lost the ability to secret immunoglobulin with a B cell known to secrete a particular antibody results in a remarkable hybrid cell that produces the antibody made by its B-cell component but retains the capacity of its myeloma component to multiply indefinitely. Britannica Academic Edition.

[11] Monoclonal antibodies:

Antibodies with a defined specificity derived from cloned cells or organisms. PRODUCTION AND QUALITY CONTROL OF MONOCLONAL ANTIBODIES, EMA.

4) Chemical medicines are well defined and need to follow specific standards, which makes it relatively easy to reproduce them if required. On the other hand, biologics are more difficult to analyze. Their production process has a high impact on the final product.

5) Related to their protein structure, biologics are usually administered by injection, whereas chemical medicines can be administered in other forms such as pills, creams, ointments and sprays, etc.

6) Due to their differences in terms of administration and the foreign nature of these particular biologically active molecules to the body, potential adverse effects (the "immunogenic response"[13]) with biologics are always a concern.

The EMA recently described biologics as:

> *"A biological medicine is a medicine that contains one or more active substances made by or derived from a biological source. Some of them may be already present in the human body and examples include proteins such as insulin, growth hormone and erythropoietins[14]. The active substances of biological medicines are larger and more complex than those of non-biological medicines. Only living organisms are able to reproduce such complexity. Their complexity as well as the way they are produced may result in a degree of variability in molecules of the same active substance, particularly in different batches of the medicine".* (2)

1.1.2 Definition of Biosimilar

In the pharmaceutical industry, the practice that *innovator pharmaceutical companies* [15] protect their inventions with patents, for obvious business reasons, is very common. After the patent expiry of the originator chemical medicine, generic versions of the originator chemical medicine are launched by competing pharmaceutical companies. The generic version of the pharmaceutical (chemical) product has the same active substance. That is possible because the active ingredients are identical to each other.

---

[12] Heterogeneous:
Adjective diverse in character or content: a large and heterogeneous collection. Oxford Online Dictionary.
[13] Immune response:
The reaction of the cells and fluids of the body to the presence of a substance which is not recognized as a constituent of the body itself. Oxford Online Dictionary.
[14] Erythropoietin:
Hormone produced largely in the kidneys that influences the rate of production of red blood cells. Britannica Academic Edition.
[15] Innovator pharmaceutical companies:
Interchangeable with the term originator pharmaceutical companies. Pharmaceutical companies which brought something new to the initial the medicinal product i.e. pharmaceutical organizations which conduct research.

Directive 2001/83/EC describes a generic as:

> *"'Generic medicinal product' shall mean a medicinal product which has the same qualitative and quantitative composition in active substances and the same pharmaceutical form as the reference medicinal product, and whose bioequivalence with the reference medicinal product has been demonstrated by appropriate bioavailability studies…."* (3)

Already, prior to its introduction, the term biosimilar was heavily disputed between regulators and the different pharmaceutical industry branches, namely the generics and innovator industries. The term biosimilar, itself, already outlines the major difference between chemical medicine generics and biosimilars as biological medicines, which are similar to an innovator biological medicine. It contains the suffix "similar", a term that means according to the Oxford Dictionary (4) *"having a resemblance in appearance, character or quantity, without being identical"*.

The innovator medicinal product is also called biological reference medicine, as it serves as a reference and standard for the biosimilar. Finally, the decision bodies within the EU agreed upon the fact that unlike standard generics, which are identical to innovator chemical medicinal products, biologics can only be similar to, but not identical to, innovator biologics.

From the regulatory perspective, the word biosimilar can be described as a "similar biological medicinal product", where "similar" is to be understood in relation to a biological reference to the medicinal product.

> *"Where a biological medicinal product which is similar to a reference biological product does not meet the conditions in the definition of generic medicinal products, owing to, in particular, differences relating to raw materials or differences in manufacturing processes of the biological medicinal product and the reference biological medicinal product, the results of appropriate pre-clinical tests or clinical trials relating to these conditions must be provided"*. (3)

Omnitrop® (Somatropin) was the first biosimilar product to be approved within the EU. In its EUROPEAN PUBLIC ASSESSMENT REPORT (EPAR)[16] the word biosimilar is described as:

> *"Omnitrop is a 'biosimilar' medicine. This means that Omnitrop is similar to a biological medicine that is already authorized in the European Union (EU) and contains the same active substance (also known as the 'reference medicine')"*. (5)

[16] EPAR: The European Medicines Agency publishes an EPAR for every medicine granted a central marketing authorization by the European Commission. EPARs are full scientific assessment reports of medicines authorized at a European Union level, source EMA.

High ranking members of the EMA gave the following definition of a biosimilar in a highly regarded scientific journal (Nature Biotechnology):

> *"A biosimilar is a copy version of an already authorized biological medicinal product with demonstrated similarity in physicochemical*[17] *characteristics, efficacy and safety, based on a comprehensive comparability exercise. As a biosimilar is highly unlikely to be identical to its reference product, the standard 'generic' approach (that is, demonstration of bioequivalence*[18] *in comparative bioavailability studies, established for small chemically derived and easily characterized molecules) is not sufficient for the development, regulatory assessment and licensing of such a product. For this reason, we argue that the term biogeneric is scientifically incorrect and should not be used for a biosimilar".* (6)

The EMA most recently described biosimilars as:

> *"A biosimilar medicine is a biological medicine that is developed to be similar to an existing biological medicine (the 'reference medicine'). Biosimilars are not the same as generics, which have simpler chemical structures and are considered to be identical to their reference medicines. The active substance of a biosimilar and its reference medicine is essentially the same biological substance, though there may be minor differences due to their complex nature and production methods. Like the reference medicine, the biosimilar has a degree of natural variability. When approved, its variability and any differences between it and its reference medicine will have been shown not to affect safety or effectiveness".* (2)

1.2 Statement of the main problem, subsequent research questions and test-functions

1.2.1 Background aspects

This section provides the basic information for understanding the research question.

The biologics life cycle includes changes to the production process, which are inevitably linked to changes to the quality profile of the biologic. As for a change in the manufacturing process, the biosimilar developer is required to show comparability to the reference medicinal product.

---

[17] Physicochemical:
Relating to physics and chemistry or to physical chemistry. Oxford Online Dictionary.

[18] Bioequivalent:
Two medicinal products containing the same active substance are considered bioequivalent if they are pharmaceutically equivalent or pharmaceutical alternatives and their bioavailability (rate and extent) after administration in the same molar dose lie within acceptable predefined limits, source GUIDELINE ON THE INVESTIGATION OF BIOEQUIVALENCE, EMEA.

However, heterogeneity of the biologics and variability of the process be it due to regulatory controlled changes or be it through unnoticed (and therefore uncontrolled) effects, result in challenges to successfully develop, conduct and accomplish a biosimilar development program.

1.2.1.1 Outline of a biologics life cycle

Generally speaking, the life-cycle of a biologic can be split into two parts, which are separated by the granting of an M.A. The development phase of a biologics innovator medicinal product includes:

- Discovery-phase, where a new compound NCE is discovered.
- Development-phases, where the format of the new medicinal product and a manufacturing process is developed and non clinical testing is carried out.
- Clinical trial phases 1, 2a, 2b and 3, where the adjustment and refinements of the development phase are tested and evaluated with regard to safety and efficacy.
- Filing/review phase through the regulator.

It can take usually between 8 to 16 years, depending on many factors. During that time frame the product is made ready for M.A. approval.

The commercial phase, after the he M.A. is obtained and the medical product is on sale for use by the patients, depends on market exclusivity and patents of the innovator medicinal products. This will be addressed in detail in section: 2.2.3 The biologics life cycle. This does not mean that after that period of time the innovator product ceases production. However, due to competition with generics and nowadays biosimilars the market share will shrink.

The exclusivity period once a patent is filed may last between 10-15 years. In other words, a biologic will be available on the market for probably dozens of years. For instance, according to Sandoz, follow-on drugs take seven to eight years to develop compared with eight to 10 years for a new drug application. (7) During that period of time, biologics may undergo many changes. Companies will be bought and sold, moves that may affect the production site will occur. Due to the demands of the market, the production might be scaled up or if facing competition might go down, and the process requires permanent adaptation.

As the supplier changes their production processes, the quality of the raw materials might change as well. The purchase of new equipment might be needed to replace whole or part of the production and this might need to be redesigned in order to answer the requests of the market. Last but not least, the production might be optimized and new technologies introduced.

From an industrial production perspective, even once the M.A. is obtained the process and therefore the product (this will be further address in section: 1.1.2 Definition of Biosimilar) remain subject to several changes. Variation is a natural fact, which from a classical point may be considered detrimental to quality. However, variation is part of the natural process and therefore unavoidable.

1.2.1.2 Requirement for a comparability exercise

Thinking about the previous definitions of biosimilars (addressed in further detail in section: 1.1.2 Definition of Biosimilar), according to the E.U. regulations EMA considers a biosimilar medicinal product as a copy of an already authorized biological medicinal product (the reference medicinal product). However, as already discussed the generic approach used for chemical medicinal products cannot be adapted for use with biologics.

Thus, any biosimilar submitted to the E.M.A. for M.A., is required to demonstrate its similarity to the reference biologic medicinal product, including a comparison of the quality attributes of the reference biologic. Those attributes are quality characteristics, efficacy and safety. Therefore, comparative nonclinical studies and usually clinical studies are required, in order to ensure close resemblance in safety and efficacy.

The summary of studies required to show the similarity between biosimilar and reference biologic is called the "comparability exercise".

- Refer also to section: 2.3.2 Guideline on comparability of medicinal products containing biotechnology derived proteins as active substance: Quality issues EMEA/CPMP/BWP/3207/00 (50) and section: 2.3.3 Guideline on comparability of medicinal products containing biotechnology-derived proteins as active substance: Non-clinical and clinical issues EMEA/CPMP/3097/02/Final (51) for more technical detail on the comparability exercise.
- Refer also to section: 2.2.3 The biologics life cycle in this study for commercial implication of the comparability exercise.

The rational for the requirement of a comparability exercise for biosimilars is mainly based on the following aspects:

1) Biologic medicinal products are composed as active substance form (larger) biological molecules such as proteins. The complexity of this molecular structure is difficult to characterize fully. In addition, from a scientific and practical point of view, the degree of heterogeneity[19] and variability of the biological system used for the manufacturing process will always impact the biological product and show variability. If variability is not detected this could mean that the assay sensitivity is not good enough (refer to section: 2.2.1 Heterogeneity of biologics (proteins only) for a discussion on heterogeneity). Simply, as already mentioned, the living cells produce proteins. This biologic process of protein manufacture requires a diverse cellular machinery of DNA, and different RNA forms, such as nucleic acids and mainly effector

---

[19] Heterogeneous:
Diverse in character or content: a large and heterogeneous collection. Oxford Online Dictionary.

proteins such as enzymes (here the keywords transcription[20] and translation[21] should be mentioned, which can be read about in many sources). Furthermore, there is also a location component, as the protein production process requires the involvement of different locations and organelles[22] within the cell. Considered together, natural heterogeneity of the protein occurs as the cellular components involved in the protein manufacturing process vary because of a wide range of factors. For example, as a response to changing environmental condition components the glycosylation[23] on the protein may change. As a result, the glycol-protein molecules, which are manufactured by the cell, are not identical (refer also to section: 2.2 Life cycle in relation to heterogeneity and variation for details).

2) Biologic medicinal products are defined through their manufacturing process. Those manufacturing processes are individual and unique for each medicinal product.

It is unrealistic to assume that a complete biotechnological process, which is usually composed of a series of steps with high complexity, can be rebuilt and reproduced. Thus, the product, i.e. the biologic, is individual and unique. The notion that the product is the process and the process is the product is completely acknowledged by EMA: refer to Annex 2 Manufacture of Biological active substances and Medicinal Products for Human Use. EudraLex Volume 4 where it states:

> *"Unlike conventional medicinal products, which are manufactured using chemical and physical techniques capable of a high degree of consistency, the manufacture of biological active substances and medicinal products involves biological processes and materials, such as cultivation of cells or extraction from living organisms. These biological processes may display inherent variability, so that the range and nature of by-products may be variable. The methods employed in the manufacture of biological active substances and biological medicinal products for human use ('biological active substances and medicinal products') are a critical factor in shaping the appropriate regulatory control. Biological active substances and medicinal products can be defined therefore largely by reference to their method of manufacture. This annex provides guidance on the full range of active substances and medicinal products defined as biological".* (8)

Due to unavoidable differences in the manufacturing process between the biologic and the biosimilar, which may include the use of different expression systems, fermentation and

---

[20] Transcription:
The process of transcribing RNA, with existing DNA serving as a template, or vice versa. Oxford Online Dictionary.

[21] Translation:
The process by which a sequence of nucleotide triplets in a messenger RNA molecule gives rise to a specific sequence of amino acids during synthesis of a polypeptide or protein. Oxford Online Dictionary.

[22] Organelle
Any of a number of organized or specialized structures within a living cell. Oxford Online Dictionary.

[23] Glycosylation:
the process by which sugars are chemically attached to proteins to form glycoproteins. The Free Dictionary, section Medical Dictionary.

purification processes, as well as different excipient, the quality attributes of the biosimilar and the reference medicinal products will not be strictly identical.

1.2.1.3 Heterogeneity and variability of biologics

Biosimilars cannot be identical to the reference biologic. There are differences to a certain extent. As both the reference biologic and the biosimilar are manufactured by dynamic and evolving processes due to raw material changes, for example, as a minor example or production site changes, variability is inherently associated with those products.

The above mentioned dogma according to which "the product is the process and the process is the product" is disputed by some experts. Evolution is viewed towards complex biologics from the theory that the source material defines the product (i.e. the process is the product) to the current thinking that the process affects the quality of the product, but does not uniquely define it (9).

Heterogeneity and variation are unavoidable facts; however, if dealt with adequately this may not be a matter of concern. This fact has been acknowledged by scientists, biologics pharmaceutical industry and also the regulatory agencies, since the introduction of biologic medicinal products.

1.2.2 Problem statement

For the biosimilar context as a whole it is important to understand well the connection with the previous and following sections as follows:

- There is the requirement to show comparability between the biosimilar candidate and the reference biologic.
- There is a life cycle of biologics, which is applicable to the innovator biologic and biosimilar.
- There is the natural variation associated with the process and inherent heterogeneity of bio-molecules, which is associated with cellular process that cannot be fully controlled.

Taking into account these three points, a biosimilar developer faces the following challenge:

To obtain its own MA for its biosimilar, they need to synchronise the biosimilar process dynamic with the unforeseeable process dynamic of the reference biologic while still meeting regulatory expectations with regard to quality, safety and efficacy during the comparability exercise.

The above thought is set into context well by McCamish et al in 2011 (10). They published a significant paper with regard to the topic.
Figure: 1 from McCamish et al in 2011 (10):

"Biosimilarity goal posts. The ‚goal posts' of biosimilarity are established by the biosimilar sponsor by their analysis of the distribution of product attributes present in the reference product pre- and post- manufacturing change. Then they use these to select the design space for their biosimilar candidate. While the complete quality range may be quite broad for the life time of the reference product, the biosimilar sponsor will select a tighter range of control for their biosimilar product".

McCamish et al describes the quality profile of the reference biologic as well as the quality profile of the biosimilar. It shows clearly that both might change and evolve over time. This is one aspect and key train of thought of this study. (The below figure was incorporated in this study and called Figure: 1.)

Figure: 1

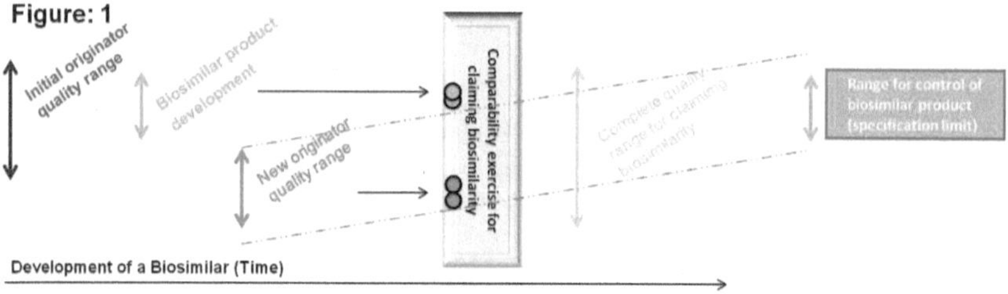

Figure: 1
The "Initial originator quality range" is the quality range of the reference biologic in dark blue.
The quality range of the quality profile of the biosimilar developer is in light blue.
Clearly, the quality profile of the biosimilar fits within the limits of the quality profile of the reference biologic.
Then in pink, the quality profile of the reference biologic has changed, whereas the quality profile of the biosimilar did not. With dissimilar quality profiles the comparability exercise is carried out.

However the conclusion that is drawn in the paper is that Figure: 1 is lacking clarity; thus, the schema should be as in Figure: 2. (11).

Figure: 2

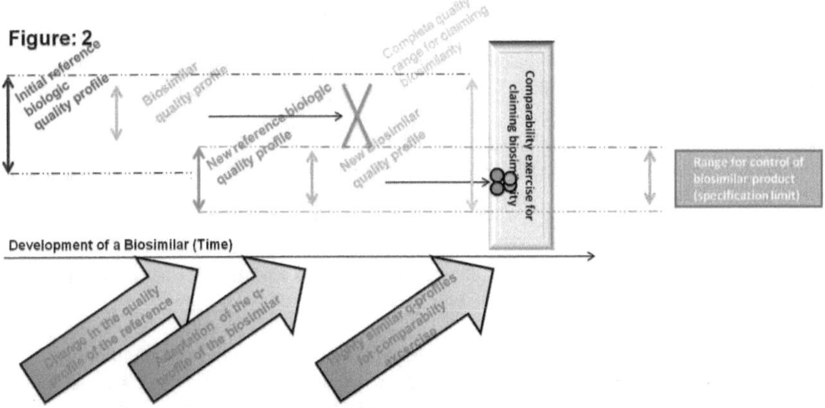

Figure: 2
The "Initial originator quality range" is the quality range of the reference biologic in dark blue.
The quality range of the quality profile of the biosimilar developer is in light blue.
Clearly, the quality profile of the biosimilar fits within the limits of the quality profile of the reference biologic.
Then in pink, the quality profile of the reference biologic has changed. The new quality profile is adopted by the biosimilar and the comparability exercise is carried out with highly similar quality profiles.
The allowable quality range (quality profile) is dictated by the quality profile of the biosimilar.

Refer also to the discussion in section: 4.4 Discussion.

Based on McCamish et al's train of thought, the research questions for this study were developed. They established correctly that the quality profiles of the reference and by consequence of the biosimilar as both are biologics are not stable. These quality profiles are dynamic when comparing them on a lot to lot basis for the same process (product) and when comparing the output of different processes (different products), as is the case for the reference biologic and the biosimilar. The heterogeneity and variation for the same process tend to be reduced compared to heterogeneity and variation for different processes.

The challenge for the biosimilar developer is to match the dynamic quality profile of the biosimilar to the dynamic quality profile of the reference biologic. Similar to the situation where in a car race one driver tries to match his speed to another car, however he is only in the position to control his own speed. Thus the matching of the speed of the two cars can be difficult, if the second driver changes his speed. That is the basic underlying concept of this study and is reflected in the research question series 1.0 to 3.0, in which the situation of the dynamic of quality profiles is evaluated and the resulting consequences of this dynamic of quality profiles are evaluated. The EMA issued a series of guidance documents on the topic that will be further address in detail in section: 2.3 Screening the above presented literature related to current biosimilar regulation with regard to the research questions. Despite the rich regulatory corpus, a great deal of questions (still) remain, as biosimilar approval is a process that is still relatively new within the EU. After a couple of years of experience, the EMA revisited many guides and clarifications were made, which also lead to the research questions below:

1.0 What are the implications of heterogeneity and variation through the life cycle of the biosimilar and the reference biologic, from a European perspective?

1.1 Taking into account the new amendment of EMA/CHMP/BWP/617111/2010, with emphasis on the life-cycle of the Biosimilar, does the Biosimilar need to achieve a quality profile which falls within the quality profile of the reference biologic?
Under which circumstances could this approach be deviated from?

1.2 What is proposed if the reference biologic changes its quality profile during the biosimilar development program?

2.0 What should be the scope of trials for biosimilars?

2.1 How extensive do biosimilar trials need to be?

2.2 Does one need to test multiple lots of the biosimilar vs. the reference medicinal product in trials?

2.3 Meta Analysis, maybe a sign for extensive trials, as such reflecting potential issues that show comparability at a first glance.

3.0 Why is extrapolation of indications for biosimilar controversial?

1.2.2.1. The dynamic of the quality profile during the life cycle of a biologic

The fundamental question within the biosimilar context is and will remain for a long time: "how similar does a biosimilar need to be to its reference biologic?" However, this is not what this study is looking to answer. For this study, the question is about exploring the issue of similarity, but only in conjunction with the whole lifecycle of the biologic. Thus, the correct question is:

1.0 What are the implications of heterogeneity and variation through the life cycle of the biosimilar and the reference biologic, from a European perspective?

The degree of heterogeneity and variation needs to be within boundaries. However, the life-cycle of any pharmaceutical product is likely to go through a certain dynamic that will be referred to further in section: 2.2.3 The biologics life cycle. Thus, what is the impact when taking those two aspects into account? A question that will be further referred to in Chapter 5.0: What should be the scope of trials?

As a direct result of research question 1.0, two concrete sub-aspects were further investigated.

One question is about the quality profile of the biosimilar. The quality profile is the summary analytical results and characterisation. The other question is related to the life-cycle and the dynamic of both the reference biologic and the biosimilar.

1.1 Taking into account the new amendment of EMA/CHMP/BWP/617111/2010, with emphasis on the life-cycle of the Biosimilar, does the Biosimilar need to achieve a quality profile which falls within the quality profile of the reference biologic? Under which circumstances could this approach be deviated from?

This issue is of relevance to this study, as this is fundamental in setting specifications for biosimilars and relates to the question of how similar a biosimilar needs to be. We will address this question in more detail in Chapter 5.0: What should be the scope of trials?

1.2 The question: "What is proposed if the reference biologic changes its quality profile during the biosimilar development program?" will be addressed in a dedicated section.

The question will be analysed with regard to the life cycle about the dynamic of the quality profile. The quality profile changes permanently. As a matter of fact, every lot is slightly different to other lots bearing a ratio of the protein isoform that is slightly different to previous lots and therefore unique. That is true for all biologics, as mentioned before.

However, the biosimilar developer usually monitors and analyses multiple reference biologic lots over extended periods, a procedure that allows for the creation of the so-called quality profile. This is a range in which the variation and heterogeneity occurs, but which is not exceeded under usual circumstances. On the other hand, as the period of time and the number of lots analysed only allows limited analysis, and the dynamic of the reference biologic is unforeseeable for the biosimilar developer, one may also encounter the situation that the quality profile of the reference biologic, as it was established initially by the biosimilar developer, changes as well. Thus making the biosimilars own quality profile different compared to the reference biologic profile.

Such an occurrence may happen at any stage of the development process of the biosimilar, a situation that poses unsolved challenges for the biosimilar developer and the regulatory authorities on how to proceed. The regulatory process is still in its infancy, as the only guidance from the regulatory side at the moment is that once the product was authorised, there are no requirements any more for similarity (refer to current mAbs and Quality guide draft: 2.3.6 Guideline on similar biological medicinal products containing biotechnology-derived proteins as active substance: Quality issues (revision 1) Draft EMA/CHMP/BWP/247713/2012 (54)).

1.2.2.2. The costs for a biosimilar

Other very relevant and problematic questions refer to the scope of trials. The costs of bringing a medicinal product to the market are of utmost interest in the industry. Therefore, regulatory aspects that impact significantly on costs, or reduces them, are of high interest. The scope and burden to show biosimilarity in clinical trials is disputed, due to their high costs. The questions on how large clinical trials for biosimilar should be are not clear. Thus a series of questions was forward to resolve this issue.

2.0 What should be the scope of trials for biosimilars?

2.1 How extensive do biosimilar trials need to be in order to show comparability?

What are the expectations and requirements based on the guidelines and the current state of discussion on clinical trials and how should a clinical trial design look like for a biosimilar clinical trial? This will reveal the requirements and current practices and will indicate the clinical trials scope including the number of subjects that participated.

2.2 Under what circumstances does one need to test multiple lots of the biosimilar vs. the reference medicinal product in trials?

Based on the dynamic of the quality profiles, does one need to take heterogeneity and variation aspects into account when conducting clinical trials, which leads to incorporation of multiple lots into the trial design? The extent on trials will evidently impact on the costs; therefore, this question has a potentially high impact on cost if answered positively.

2.3 Meta Analysis, maybe a sign for extensive trials, as such reflecting potential issues that show comparability at a first glance.

Meta Analysis can be interpreted as a sign that clinical comparability data was not conclusive enough or it has been controversial that questions the comparability. Thus, this will indicate the suitability of the clinical trial design employed, including its scope.

3.0 Why is extrapolation of indications for biosimilar controversial?

The EMA proposed to extrapolate indications without the requirement of additional clinical trials under the condition that for one indication biosimilarity was established and the mechanism of action is identical.

1.3 Study "Within Scope" and "Out of Scope"

This study deals with European biosimilars and more particularly analyses the framework set by the following guidance documents as questions and comments were send to the EMA with regard to those guidance documents, refer to Appendix: 1 Questions and comments submitted to the EMA and Appendix: 2 Questions and comments submitted to the EMA.

- GUIDELINE ON SIMILAR BIOLOGICAL MEDICINAL PRODUCTS CONTAINING BIOTECHNOLOGY-DERIVED PROTEINS AS ACTIVE SUBSTANCE: NON-CLINICAL AND CLINICAL ISSUES
  EMEA/CHMP/BMWP/42832/2005
  http://www.ema.europa.eu/docs/en_GB/document_library/Scientific_guideline/2009/09/WC500003920.pdf

- Guideline on similar biological medicinal products containing biotechnology-derived proteins as active substance: quality issues (revision 1) **Draft**
  EMA/CHMP/BWP/247713/2012
  http://www.ema.europa.eu/docs/en_GB/document_library/Scientific_guideline/2012/05/WC500127960.pdf

- Guideline on similar biological medicinal products containing monoclonal antibodies **Draft**
  EMA/CHMP/BMWP/403543/2010
  http://www.ema.europa.eu/docs/en_GB/document_library/Scientific_guideline/2010/11/WC500099361.pdf

- Guideline on similar biological medicinal products containing monoclonal antibodies – non-clinical and clinical issues
  EMA/CHMP/BMWP/403543/2010
  http://www.ema.europa.eu/docs/en_GB/document_library/Scientific_guideline/2012/06/WC500128686.pdf

The main focus is on the implications of heterogeneity and variation through the life cycle of the biosimilar and the reference biologic, as stated in the main research questions.

The follow up theme of this main research question is the cost to conduct a comparability exercise based on regulatory exceptions.

This study does not envisage evaluating any situation external to the EFTA/EU regulatory situation, although a brief description of the biosimilar situation and links in the US, Canada and Japan will be mentioned for informational purposes in section: 2.4. Reference to other biosimilar regulations.

1.4 Description of the European Regulatory Environment with Regard to Biosimilars

The European Medicines Agency EMA (previously on called EMEA) was founded in 1995 in accordance with the European Community Regulation (ECC) No 2309/93. The EMA is responsible for the approval, assessment, supervision and monitoring of the medical domain and to promote public health in the EU/EEA, respectively the EFTA countries.

The EMA relies on the scientific resources of the national drug agencies (competent authorities in the respective 30 member states of the European Union and EEA countries). Also, its personal often play a decision making role within the national authority, but also within the EMA.

Through 2001/83/EC, via the so-called centralized procedure, formed a committee, the CHMP (for human medicines), to give scientific advice. CHMP can seek advice from the biologics working party (BWP). The BWP provides recommendations to the EMA scientific committees, such as the CHMP, on all matters relating directly or indirectly to quality and safety aspects relating to biological and biotechnological medicinal products. This includes biosimilars.

When it comes to granting a Marketing Authorisation (MA), the European Commission usually follows the recommendation of the CHMP.

The legal basis under which the EMA exercises its duties is the community regulation (EC) No 726/2004. The most important of the main regulations (directives and regulations) for medicinal products is undisputedly Directive 2001/83/EC that will be referred to later in section: 2.3 Screening the above presented literature related to current biosimilar regulation with regard to the research questions. Amendments of this Directive 2001/83/EC by European Directives in 2003/63/EC and 2004/27/EC, gave a new legal framework for biosimilars in Europe.

Article 10 of 2001/83/EC was changed in order to require the highest standards of safety and efficacy for biologics clinical data. This laid down the foundation for all subsequent guidance documents on how to interpret and conduct the comparative exercises (12).

1.5. Why are biosimilars interesting for the generic industry?

The main reason for industrial interest in biosimilars is financial. The biosimilar has lower development costs, will be approved following a faster timeline and there is an established market where the product is accepted.

> *"Worldwide sales of biologic drugs exceeded US$92 billion in 2009. With many biopharmaceutical patents expiring over the next decade, a wave of second-generation or 'follow-on' biologics will be vying for market share and regulatory approval. Patents cover not only the drugs, but also the molecular modalities that facilitate their high-level expression".* (13)

Innovator biologics are highly evaluated within the categories of biologics. For 2013, the volume of biologics sales is estimated to reach the level of 100 million EUR worldwide. As patents on many of those biologics are expiring this raises significant opportunities for biosimilar developers.

> *"Patents on several biopharmaceuticals have recently expired, or are due to expire".* (14)

Whereas the classical division between the generics industry and innovator industry does not exist, all stakeholders try to gain market share from competing companies. The hurdles to enter the biosimilar market are significantly higher compared to classical generics; however, there are also potentially high profit margins. This also attracts pharmaceutical companies as well as the generics giants.

The classical generic companies have difficulties in entering the biosimilars market. Therefore, the potential companies that may have the resources and experience to access this more complicated biosimilar market may be different.

For example, the mAbs are the largest and fastest growing category of biologics. Over 30% of new medicinal products, with an increasing tendency, are based on biologics. A very big portion of that is mAbs. Patents on these mAbs are expiring and many biosimilar versions are currently in different phases of development or are about to be submitted for approval via the centralized procedure to the CHMP. The cost for developing a suitably complex biosimilar, for example a mAb, could be estimated to be between 80 million EUR and 250 million EUR. The cost for a new biological medicinal product could be in the order of 1 billion EUR or more.

> *"The pharmaceutical industry notes that the steep cost of developing an innovator biologic ($1.2 billion) and the lengthy time it takes a manufacturer to jump through the research, clinical testing and approval process (roughly ten years) warrants strong patent and patient protections. The American Association of Retired People (AARP) and other advocacy groups challenge those R&D figures. But there is no question that developing these treatments is costly, with many failures leading to a relative handful of well-*

rewarded successes. For better or worse, time and expense are the reality of the drug development model in a tightly regulated environment". (15)

On Sandoz's website, the company states that a biosimilar can cost between $75 million and $250 million to develop, with only 500 patients in clinical trials. (7) Some experts even estimate that costs could be lower. (16)

Novartis through its generics specialist Sandoz, Teva in combination with Lonza and Hospira are likely to be the leading players in forming and creating biosimilar versions of the blockbuster antibodies. However, the market entry cost is relatively high with at least 80million EUR needed.

"The research group Collins Stewart has estimated that developers will need to budget $100 million for the kinds of clinical trials that will be required to gain an approval. And once they hit the market, the follow-on are expected to offer discounts of 10 to 15 percent". (17)

"In the European Union, Sandoz Inc. has successfully marketed three biosimilar drugs along with its biologics pipeline. In December of 2011, Baxter International Inc., based in Deerfield, Ill., announced collaboration with Momenta Pharmaceuticals Inc. to develop six biosimilars, or follow-on biologic products. Amgen Inc. and Watson Pharmaceuticals Inc. announced Dec. 19, 2011 that the two companies will draw upon their respective strengths to develop and market biosimilars in the specialty and generic markets. Watson Pharmaceuticals is expected to pay up to $400 million in costs over the development and marketing of several biosimilar products. In addition, Celerion and Ricerca Biosciences announced the formation of "The Biosimilars Alliance" in February of 2012. Celerion stated predicted the biosimilar market will grow in the U.S. from $2.4 billion in 2012 to $44 billion by 2020". (18)

Pfizer's new pact with Biocon on biosimilar insulin products will help drive development of far more complex biologics, including the first round of biosimilar antibodies around 2014 and 2015.

1.5.1 What is the market size? What are the growth estimates for biologics and biosimilars?

Biosimilars are the future; this view is not only shared by the representatives of the biosimilar industry, but also by the classical innovator industry, where some players have entered the field of biosimilars as well.

The EMA has introduced regulatory frameworks for the approval of biosimilar mAbs medicinal products. However this path relies on a comparability exercise as outlined in Chapter 5.0: What should be the scope of trials? Thus, contrary to the classical generics pathway for M.A. approval, for biosimilars clinical trials are required, which are a significant cost factor, refer to section 5.3.1.2 Clinical trials conducted in the EU or USA. Total clinical trial requirements are difficult to foresee and are data driven, thus it will be a case by case decision that will be addressed in detail in Chapter 5.0: What should be the scope of trials?. mAb are one of the important categories of biologics, and for this study the focus will mainly be on mAbs.

Worldwide biologics sales in 2009 were at ca. 93 billion dollars and are expected to continue to grow at least twice as fast as those of small molecules. In 2010 sales of biologics amounted to 30 billion dollars in the United States, and EUR 60 billion in Europe (19). For example, many new and future biopharmaceuticals (biologics) are biotechnology derived mAbs. Ca. 300 (mAbs) are being developed (2011) for over 200 indications for oncology, inflammatory diseases, autoimmune diseases, metabolic and central nervous system disorders, infectious and cardiovascular diseases. (20)

Many blockbuster biopharmaceutical patents are expected to expire over the next decade. Thus, there is a good opportunity for the biosimilar industry to gain access and market share of this emerging biologics market with ever growing revenues. (21)

Biopharmaceuticals usually cost much more per patient than conventional pharmaceuticals, and their use is growing at a much higher rate than that of the overall pharmaceutical market. Annual treatment cost can span between 37,000 to 200,000 USD per year. (21)

It is estimated that in 2010, some 50% of all newly approved medicinal products will be biologics.

> *"There are now roughly 500 new biopharmaceutical products in various stages of development worldwide, 300 of them in Europe. In 2003 for example 7 of the 50 best selling medicines in Germany were biopharmaceuticals".* (22)

There are 150 marketed biologic products worldwide, with over 370 additional products under development (16). Worldwide biologic sales exceeded 100 billion USD in 2011, and 32% of those sales were related to mAbs (23).

> *"Further it is anticipated that in 2016, biologics half of the topselling biologics; of these, seven (Humira®, Avastin®, Rituxan®, Herceptin®, Remicade®, Prolia® and Lucentis®) are mAbs and one (Enbrel®) is a fusion protein containing antibody components."* (10)

In the top 10 best-selling drugs there will be as many as 8 biological drugs by 2014 (Table 1) compared to six in 2010 (Table 2), five in 2008 and only a single entry in 2000. In 2014 biologic drugs will account for 75% of sales, a two-fold increase in the relative contribution to the pharmaceutical sector, it has been estimated. They will mainly be used to treat cancer and rheumatoid arthritis.

The situation is predicted to change dramatically by 2014, see Table 2. At least six, and as many as eight, of the top 10 best-selling drugs are expected to be biological drugs. Moreover, biological drugs are predicted to account for 75% of sales, an almost two-fold increase in the relative contributions of biological drugs to the pharmaceutical sector.

Table 1:
Predicted top 10 best-selling medicines 2014. (24)

Rank	Brand name	Generic name	Company	Treatment	Sales 2014 (US$ bn)
1	Avastin	bevacizumab	Roche	Cancer	9.2
2	Humira	adalimumab	Abbott	Arthritis	9.1
3	Rituxan	rituximab	Roche	Cancer	7.8
4	Enbrel	etanercept	Pfizer (Wyeth)	Arthritis	6.6
5	Lantus	insulin glargine	sanofi-aventis	Diabetes	6.4
6	Herceptin	trastuzumab	Roche	Cancer	5.8
7	Crestor	rosuvastatin	AstraZeneca	Cholesterol lowering	5.7
8	Spiriva	tiotropium	Boehringer Ingelheim	COPD	5.6
9	Remicade	infliximab	J&J/Merck	Arthritis	5.2
10	Gleevec/Glivec	imatinib	Novartis	Cancer	5.1

Source: Evaluate Pharma
Bn: billion, COPD: chronic obstructive pulmonary disease; J&J: Johnson & Johnson.
: biological drug

Table 2:
Top 10 best-selling medicines 2010. (24)

Rank	Brand name	Generic name	Company	Treatment	Sales 2010 (US$ bn)
1	Lipitor	atorvastatin	Pfizer	Cholesterol lowering	11.7
2	Plavix	clopidrogel	Bristol Myers Squibb/sanofi-aventis	Anti-clotting	9.63
3	Advair	fluticasone/ salmeterol	Glaxo SmithKline	Asthma, COPD	9.0
4	Remicade	infliximab	Merck/J&J	Arthritis	7.4
5	Enbrel	etanercept	Pfizer (Wyeth)	Arthritis	7.1
6	Humira	adalimumab	Abbott	Arthritis	6.8
7	Avastin	bevacizumab	Roche	Cancer	6.7
8	Rituxan	rituximab	Roche	Cancer	6.1
9	Diovan	valsartan	Novartis	Hypertension	6.0
10	Crestor	rosuvastatin	AstraZeneca	Cholesterol lowering	5.8

Source: Times of India, 30 November 2010
Bn: billion; COPD: chronic obstructive pulmonary disease; J&J: Johnson & Johnson.
[shaded] : biological drug

Many small molecule drugs will be off patent in 2014, such as Lipitor (atorvastatin), Plavix (clopidogrel), Advair (fluticasone/salmeterol) and Diovan (valsartan), all current best sellers, and there will be major competition in the generics market due to this. In addition, from 2011 to 2015 biosimilars with a combined worth of US$17 billion will lose patent protection. (24)

It was estimated that biosimilar sales worldwide in 2011 were ca. $16.4 billion (25). The global biosimilars market is expected to grow to $17.9 billion by 2017, according to a press release issued by Global Industry Analysts Inc. However, there are also risks involved in the biosimilar market, which only the bigger players are willing to take or are capable of managing through their bigger financial means and expertise.

Clearly it is with worth citing and outlining those risks, as they highlight potential rift lines within the regulations and disputed areas.

Proteins, as the basis of most biologics, are not simple to manufacture, as outlined further in section: 2.2.3 The biologics life cycle, leading to higher costs related to manufacturing. They also require product safety that is higher than for small molecule[24] generics due to their

[24] Molecule:
A group of atoms bonded together, representing the smallest fundamental unit of a chemical compound that can take part in a chemical reaction. Oxford Online Dictionary.

potential increased immunogenicity[25]. Therefore, an extra burden of evidence is required for their approval, which means more time and more money to demonstrate product safety, as well as facing greater competition.

On the other hand, the degree of success of biosimilars is influenced by many factors. One of the most important is represented by the regulatory environment, the topic analysed in this study.

Taking again the example of mAbs, biosimilar mAbs constitutes a unique class of biosimilars. mAbs are heterodimeric[26] proteins, which are glycosylated. The (stringent) regulatory guidance for biosimilar monoclonal antibodies results from the complexity of their structure. Currently, there are still no mAbs approved within the EU, but several are awaiting approval.

It is expected that in the near future, biosimilar mAbs will be approved; however, some hurdles could infringe their economic success:

i) In view of their complexity the development costs can be expected to be significantly higher compared to that of "regular" biosimilars. Consequently, price reductions compared to the reference may be very small.
ii) Biosimilar MAs are also expected to experience significant competition from the next-generation MAs that will exhibit improved properties, e.g., through glycoengineering.
iii) Antibody-based therapeutics, such as Fc-linked fusion proteins[27] exhibiting improved therapeutic potential and improved pharmacodynamics[28], may also compete with biosimilar MAs.
v) Other antibody formats currently under development such as nanobodies[29], produced in bacteria and thus expected to be much cheaper may well form an alternative for reducing costs in healthcare (26).

---

[25] Immunogenic:
Relating to or denoting substances able to produce an immune response: immunogenic vaccines. Oxford Online Dictionary.

[26] Hetreodimer:
Hetero – Prefix, the same.
Dimer - A molecule or molecular complex consisting of two identical molecules linked together. Oxford Online Dictionary.

[27] Fc-linked fusion proteins:
Fc-based fusion proteins are composed of an immunoglobin Fc domain that is directly linked to another peptide. (102)

[28] Pharmacodynamics:
The branch of pharmacology concerned with the effects of drugs and the mechanism of their action. Oxford Online Dictionary.

[29] Nanobodies:
Are antibody-derived therapeutic proteins that contain the unique structural and functional properties of naturally-occurring heavy-chain antibodies. http://www.ablynx.com/en/research-development/nanobody-technology/understanding-nanobodies/

Another issue is represented by potential market saturation, where many companies manufacture biosimilars for the same reference biologic and if that occurs, *"it is going to be more difficult to recoup your investment dollars"* (18).

1.5.2 Patent protection and market exclusivity

Patent protection and market exclusivity is a key hurdle for biosimilar approval. However, those barriers are beginning to disappear, as they are time limited. For example, Sandoz estimates that innovator biologics worth $63 billion in sales will lose patent protection by 2015, which represents 40% of the total biologic market. (18)

This evaluation is largely in line with other estimates from consultancy experts in the field of biosimilars, which estimate that between 2009 and 2019, 21 blockbuster biologics with sales of over 50 billion dollars will lose patent protection. These are biologics mainly in the areas of oncology, inflammatory and cardiovascular diseases that will most likely lose their patent protection, which are often based on mAbs. (16)

Therefore, biosimilars remain an attractive investment project compared to the lengthy pathway from discovery to approval of a biologic medicine, which is long and uncertain, and can consume enormous resources and time. As for one successful innovator biologic between 10 and 15 years and costs on average of $1 .2 billion can be envisaged.

Innovator biologics are usually protected by an exclusivity phase (from the regulator, which is 12 years from the date of granting the M.A., for example) and also by a patent phase which is usually 20 years. The innovator industry usually tries to protect the source of their revenue by patents if feasible. The available exclusivity time frame on the market is usually limited by the patent protection time frame as it is the longer lasting one.

The patent is normally filed when the clinical trial phase begins, which can take up to 3 or 6 years. Given that some time is required to file for an M.A. with the regulator, the remaining market exclusivity timeframe is usually 13 to 16 years, given that some time will elapse for the regulatory approval process.

During the protected timeframe of the innovator pharmaceutical product, the innovator companies try to maximize their revenues to compensate for the expenses during the development phase.

The different types of protection of the innovator biologic medicinal product have to be taken into account by biosimilar developers, these are:

- Patents (20 years protection)
- Patent term extensions (plus 5 years protection)
- Market exclusivity (usually 12 years)

"Patents in the US and Europe last 20 years". Because the patent generally lasts longer than the exclusivity period granted for a New Chemical Entity (NCE) by the regulatory agencies the patent is often the barrier to generic entry rather than the exclusivity period. However, since the drug development and regulatory process can often be lengthy, both the US and Europe provide exclusivity extensions for patented drugs in order to compensate somewhat for some of this delay. The relationship between the date of patent filing and the initiation of clinical trials and the relationship between the date of patent filing and the market authorization date on the expected duration of marketing exclusivity are discussed in the following series of two articles". (27)

Market exclusivity is the time period during which the new medicinal products are protected from direct competition from generic versions, which includes biosimilars.

"Every day of market exclusivity is a potential profit for an originator company because generic drug companies capture 80% market share within six months of entering the US market". (27)

"The FDA grants for a limited period of time market exclusivity for each newly approved drug or formulation:
5 years for a new chemical entity (NCE)
3 years for a new formulation of an existing drug
7 years for an orphan drug" (28)

"After 2005 the EMA grants for a limited period of time market exclusivity in the following way:
8 years of data exclusivity dating from the EMA authorisation decision (before that, no generic applications may be filed)
Plus 2 years extra, for marketing protection: no generic applications may be approved.
Plus 1 year extra, for new indication(s) if it constitutes a significant clinical benefit". (27)

However, those periods can be extended:

- *"The US Patent Term Extension (21 CFR 60 and 35 USC 156) allows for up to 14 years of market exclusivity*
- *The European Supplemental Protection Certificate (SPC; described in EEC Council Regulation No. 1768/92) provides extensions for up to a maximum of 15 years of market exclusivity".* (28)

Patents may also create effective protection. However, there are often disputes surrounding them. Infringement on a claimed invention of a granted patent usually results in legal disputes. Patents usually last longer than the period of market exclusivity granted by the regulator.

"For patent applications that were filed after June 1995, statutory patent expiry in the United States and Europe occurs 20 years after the date of filing the non-provisional patent application (for example, a PCT (Patent Cooperation Treaty) application).

A provisional patent application is often filed 1 year earlier; in this case, patent expiry would occur 21 years after the provisional application was filed. However, because the patent holder does not receive commercial benefit until the product has been approved by regulatory agencies, authorities in the United States and Europe both provide exclusivity extensions for patented drugs, which are intended to compensate for patent life that is lost during the lengthy drug development and regulatory process. Such extensions can prolong the effective patent expiry date beyond 21 years after the filing of the provisional application". (28)

However, extensions of the patents are possible. Within the EU a supplementary protection certificate (SPC) can be granted after the patent expires. This SPC grants another 5 years to the standard of 20 years granted per patent No 469/2009/EC. In the case where data from paediatric clinical trials is added to the dossier, another 0.5 years are grated for that, as set out in Article 36 of Regulation 1901/2006/EC. The total additional protection period based on the SPC can exceed 5.5 years for human medicinal products. The combined period of market exclusivity based on a patent and SPC should not exceed 15.5 years.

The FDA, for example, allows for a patent extension of up to 5 years according to the regulations governing the Patent Term Restoration program, which are located in the Code of Federal Regulations, 21 CFR Part 60. However, the maximum of 5 years can only be granted if a maximum of 14 years from the biologics approval date (M.A.) would not be exceeded. (29)

Chapter 2: Literature review

The literature research in this chapter is subdivided into the following sections:

2.1. Re-presenting the current biosimilar legislation and regulatory requirements

(This includes a reference to section: 4.2.2. Questions and comments submitted to the EMA, where comments were forwarded to the EMA in relation to the mAbs guide in 2011.)

The literature published online was used exclusively for this study, as there was no access possible to D.I.T. or other libraries from aboard. For the elaboration of this study, literature available exclusively online was employed and a multilayer literature review was conducted where dedicated keywords were employed and were carefully selected to obtain specific information on the research questions. Whenever possible, the reference papers and articles were downloaded, also refer to the
Bibliography or References. The main sources consulted were the following:

1) EMA website, including EMA guidance documents on biosimilars (30)
http://www.ema.europa.eu/ema/index.jsp?curl=pages/regulation/general/general_content_000408.jsp&mid=WC0b01ac058002958c
This site contained all the relevant guidance documents related to the stages of approval.
2) Pubmed (31)
http://www.ncbi.nlm.nih.gov/pubmed/
The Pubmed online library contains regularly published articles specifically related to the research questions. However, the authors of those articles may have diverse links to industry thorough their research field of biosimilars so caution and critical thinking are strongly recommended.
3) Gabionline (32)
http://www.gabionline.net/
Gabi-online is associated mainly with the generics and biosimilar industry, from where it presumably receives its funding (Pro Pharma Communications International among others).
4) Google (33)
www.google.com/
For very general purposes, the main search engine used was Google; it was not used to tackle the research questions.

2.2 Life cycle in relation to heterogeneity and variation

2.3 Screening the above presented literature related to current biosimilar regulation with regard to the research questions

1.0 What are the implications of heterogeneity and variation through the life cycle of the biosimilar and the reference biologic, from a European perspective?

1.1 Taking into account the new amendment of EMA/CHMP/BWP/617111/2010, with emphasis on the life-cycle of the Biosimilar, does the Biosimilar need to achieve a quality profile which falls within the quality profile of the reference biologic?
Under which circumstances could this approach be deviated from?

1.2 What is proposed if the reference biologic changes its quality profile during the biosimilar development program?

2.0 What should be the scope of trials for biosimilars?

2.1 How extensive do biosimilar trials need to be?

2.2 Does one need to test multiple lots of the biosimilar vs. the reference medicinal product in trials?

2.3 Meta Analysis, maybe a sign for extensive trials, as such reflecting potential issues that show comparability at a first glance.

3.0 Why is extrapolation of indications for biosimilar controversial?

2.4. Reference to other biosimilar regulations

- US
- Japan
- Canada
- WHO

2.1. Re-presenting the current biosimilar legislation and regulatory requirements

2.1.1 The European biosimilar approval pathway and its regulatory framework

The regulatory framework:

All medicinal products manufactured using biotechnological processes, including biosimilars, are required to be approved via the "Centralized Procedure". For biosimilars, the European Commission amended article 10 of directive 2001/83/EC in 2004 and opened thereby a new path to biosimilar approval. The amendment was carried out through directive 2004/27/EC.

Directive 2004/27/EC displaced the term essential similarity with two new terms: generic medicinal products and similar biological medicinal product (3).

Directive Article 10(4) 2001/83/EC and Section 4, Part II, Annex I of the Directive 2001/83/EC lay down the requirements for the Marketing Authorisation on a legal basis. The first guidance specific for biosimilars was issued from 2004 onwards; whereas, already in 2003, biosimilars were included in guidance documents for comparability exercises. It is outlined that the CHMP[30] will issue specific guidance with regard to data requirements, substantiate and proof of similarity. Similarity of the biosimilar to the reference product is the basis for the M.A. application for a biosimilar medicinal product.

Article 10(4) of the Directive 2001/83/EC states:

> *"Where a biological medicinal product which is similar to a reference biological product does not meet the conditions in the definition of generic medicinal products, owing to, in particular, differences relating to raw materials or differences in manufacturing processes of the biological medicinal product and the reference biological medicinal product, the results of appropriate pre-clinical tests or clinical trials relating to these conditions must be provided. The type and quantity of supplementary data to be provided must comply with the relevant criteria stated in the Annex and the related detailed guidelines. The results of other tests and trials from the reference medicinal product's dossier shall not be provided."*

The biosimilar submission and approval process:

The CHMP is responsible for preparing the Agency's opinions on all questions concerning medicines for human use within the European Union, in accordance with Regulation EC 726/2004. Its responsibilities include conducting the initial assessment of medicines for authorisation, the post-authorisation and maintenance activities as well as modifications and extensions to an existing marketing authorisation. The EMAs scientific committee (usually CHMP for human medicinal products) designates a rapporteur and a co-rapporteur who receive the full copies of the marketing authorisation application file.

The rapporteur plays the key role in this process even after the authorisation is granted. Their main tasks are: assessing and submitting the applicant replies for discussion, as well as preparing a final assessment report. This process ensures that the respective medicines have a positive risk-benefit balance in favour of patient benefits.

---

[30] CHMP:
The Committee for Medicinal Products for Human Use (CHMP) is the committee at the European Medicines Agency that is responsible for preparing opinions on questions concerning medicines for human use.
http://www.emea.europa.eu/ema/index.jsp?curl=pages/about_us/general/general_content_000094.jsp&mid=WC0b01ac0580028c79

The CHMP provides a positive or negative opinion on the product, as well as the granting of the authorisation by submitting a draft summary of the product characteristics, the package leaflets and the texts proposed for the various packaging materials. This process takes at most 210 days after which EMA will have to forward its opinion to the Commission.

In a second phase, the Commission ensures that the marketing authorisation is compliant with the EU directive 2001/83/EC from 2004 (latest version). The EMA sends to the Commission its opinion and assessment report together with annexes containing the summary of the product characteristics (Annex 1), the details for batch release, the details and the manufacturer of the biological active substance as well as the condition of the market authorisation (Annex 2), the labelling and the packaging leaflet (Annex 3).

There is a delay of fifteen days during which the Commission has to prepare a draft decision. In the end, the medicinal product will obtain a Community registration number that will be placed on the packaging in the case where the marketing authorisation is granted.

In cases where a favourable opinion is obtained, then there will take place the empowerment procedure by the Commission's Secretariat-General who notifies the Commission decision to the marketing authorisation holder and furthermore publishes it in the Community register. The marketing authorisations are valid for 5 years and the applications for renewal are requested to be made to the EMEA at least 6 month before the 5-year period expires.

The CHMP publishes a European Public Assessment Report (EPAR) for every centrally authorised medicine that is granted a marketing authorisation, setting out the scientific grounds for the Committee's opinion in favour of granting the authorisation, plus a 'summary of product characteristics' (SPC), labelling and package leaflets (most specifically, the patient/user information leaflet) for the medicine, and details of the procedural steps taken during the assessment process. EPARs are published on the Agency's website, and are generally available in all official languages of the EU.

In 2006 Omnitrop[31] received its M.A. and was the first biosimilar approved within the EU, it is manufactured by SANDOZ (34, 35).

2.1.2 Regulatory guidance literature

General literature on biosimilars can be found on the EMA website (36). Summary of the main items has been included in Figure: 3.

---

[31] Omnitrope:
Generic Name – somatropin
Omnitrope is a form of human growth hormone. http://www.drugs.com/omnitrope.html

The literature research was limited to:

1. Guidance documents about the principles of biosimilarity.
2. General guidance documents for quality and clinical issues.
3. Product specific guidance documents for mAbs, as the most recent product specific guidance available from the EMA.

The reasons:

1. Larger molecules with a certain complexity represent the most challenging regulatory topic.
2. For some of those guidance documents comments could be forwarded to EMA.
3. Assessing all available guidance documents would represent a too big challenge for the scope of this study, especially taking into account time restrictions.

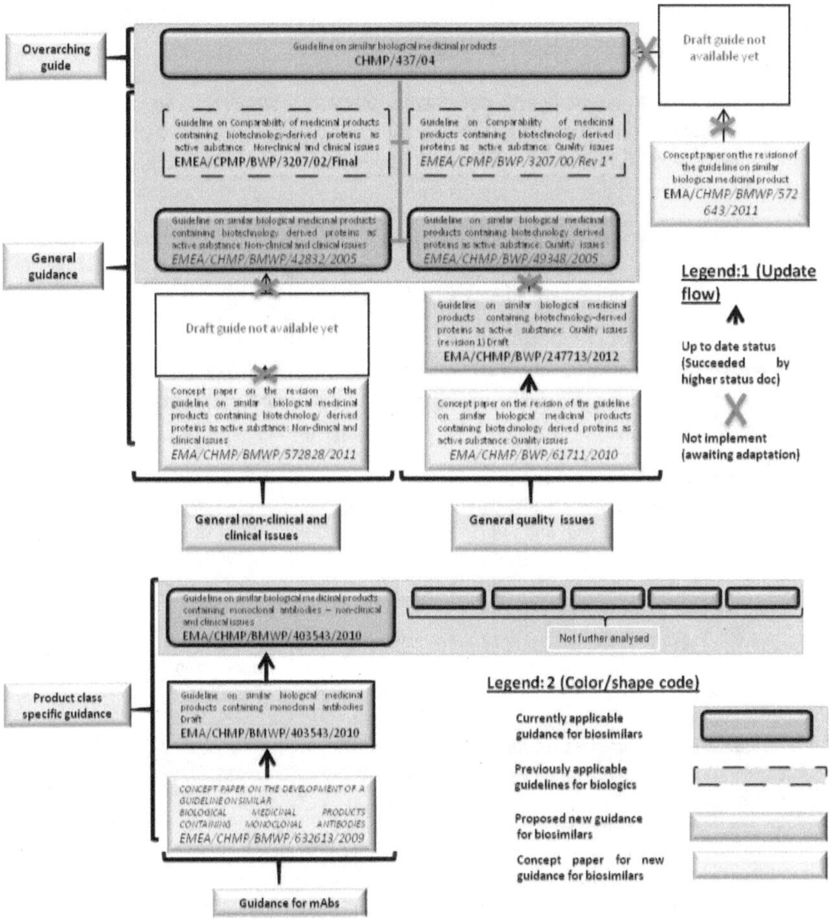

Figure : 3
Overview of the regulatory guidance literature used as the basis of this study, including the overarching guides and the general guides.
Product specific guides are not further evaluated, with exception of the mAb guide, as part of Appendix: 1.

2.1.3. Review of the state of regulation prior to the submission of questions and comments in relation to two draft biosimilar guidance documents

For the following proposed or old documents, questions, comments and suggestions for improvement were submitted to the EMA. The submission process is open to the public and the draft guidance itself contains the information about how to proceed in the case where it is desired to submit questions and comments with regard to the draft, to the EMA.

1. A word format template is proposed for the submission.
2. A postal address is provided.
3. Equally final submission deadlines were indicated.

There was the opportunity to submit comments to the EMA until the end of 2011 for the draft guide on the guideline on similar biological medicinal products containing monoclonal antibodies:

1. **The proposed draft guidance:**
   "EMA/CHMP/BMWP/403543/2010 Guideline on similar biological medicinal products containing monoclonal antibodies Draft"

2. **The following comments were made by M. Osmane in relation to the research questions treated in this study:**
   EMA/205886/2012 Overview of comments received on 'Guideline on similar biological medicinal products containing monoclonal antibodies' (EMA/CHMP/BMWP/403543/2010), also referred to
   Appendix: 2
   Questions and comments submitted to the EMA

3. **The outcome:**
   The outcome is further analyzed as part of chapter 4, section: 4.2.3 Survey research questions of this study and the responses from the EMA are available under:
   "EMA/CHMP/BMWP/403543/2010 Guideline on similar biological medicinal products containing monoclonal antibodies – non-clinical and clinical issues".

There was a further opportunity to submit comments to the EMA till the end of 2012 for a revised guide on similar biological medicinal products containing biotechnology-derived proteins as active substance: quality issues.

1. **The proposed draft guidance:**
   EMA/CHMP/BWP/247713/2012
   Guideline on similar biological medicinal products containing biotechnology-derived

proteins as active substance: quality issues (revision 1)

2. **The following comments were made by M. Osmane in relation to the research questions treated in this study:**
   Mentioned in Appendix: 2.

3. **The outcome:**
   The outcome will be further analyzed as part of chapter 4, section: 4.2.3 Survey research questions of this study.
   At the time of submission for this study, the EMA had not answered the submitted comments.

2.2 Life cycle in relation to heterogeneity and variation

Biologics based on recombinant[32] biotechnology are usually proteins; therefore, their size can range from roughly 19kDa (for interferon-alpha) to roughly 150 kDA (for IgG molecules); they have a complex three dimensional structure, which allows them to exert their biological functions.

When proteins interact in the body, the mode as well as the chemical interaction (bonding to a target molecule, docking to a receptor, etc. is based on various types of chemical interaction, such as hydrogen bonds, hydrophobic interactions[33], Van der Waals[34] and ionic bonds[35] that can trigger a multitude of biological effects).

Nevertheless the complexity of proteins is not limited to the above basic considerations. In the same way that nature does not like empty spaces, or strict mathematical shapes, proteins, as one of nature's products, are imperfect in a mathematical (or chemical) sense. They exhibit a natural variability.

In this study the term 'heterogeneity' is employed for describing this aspect. It means that there are always differences between such large and complex molecules. Surely, those

[32] Recombinant:
Relating to or denoting an organism, cell, or genetic material formed by recombination: the DNA of these recombinant viruses revealed no signs of instability, according to Oxford Online Dictionary.
[33] Hydrophobic interactions:
Another type of attraction is that between nonpolar side chains of valine, leucine, isoleucine and phenylalanine; the attraction results in the displacement of water molecules and is called hydrophobic interaction. Britannica Academic Edition.
[34] Van der Waals forces:
Weak, short-range electrostatic attractive forces between uncharged molecules, arising from the interaction of permanent or transient electric dipole moments, according to Oxford Online Dictionary.
[35] Ionic bond:
Also called electrovalent bond, type of linkage from the electrostatic attraction between oppositely charged ions in a chemical compound. Britannica Academic Edition.

difference often have no effect or are minor; however, they do exist, are present and will be always be present, assuming of course that the detection method is detailed enough to detect them. Heterogeneity is therefore the result of the bioprocesses that lead to the construction, modification and destruction of protein. The biopharmaceutical can be produced as different isomers and can form aggregates. The structure and activity of some proteins are complicated by glycosylation and other posttranslational modifications that lead to heterogeneity. (37)

In contrast to this situation, there is variation that is meant as a summary of variability introduced by man-made technology. This starts with upstream processes such as cell banking, the conditions of the bioreactor[36] in which the cells are manufactured, the protein of interest is multiplied to the downstream processes of harvesting the target protein and separation of it via a series of purification steps usually accompanied by various filtration and virus inactivation steps.

All of those steps may potentially impact on the variability of the biologic due to technical influences of the production process. For example, the purification process will highly impact which isoform(s) of the manufactured target protein will be finally present in the active substance. Therefore, the quality profile of the biologic is highly dependent of the production process, in line with Annex II of EudraLex Vol. 4[37] 'the process is the product'. (38)

2.2.1 Heterogeneity of biologics (proteins only)

If biologics are considered to be proteins only, then the following sources of heterogeneity can occur. This takes into consideration the previous definition of heterogeneity, as a biological variation caused by nature. However for simplification purposes, we define heterogeneity within the limitations of a single cell.

Errors can occur at the DNA level. For example, external factors such as UV light and radiation can cause DNA damage to coding regions. Also the DNA replication enzyme (DNA-polymerase) has a certain error rate and inserts the wrong nucleotide or too many or too few nucleotides into a sequence. Various repair processes fix most errors, but some errors become permanent mutations in the coding region. If those issues occur, they are passed on to subsequent cellular generations. (39)

Then the DNA gets transcribed into mRNA in the nucleus. Point mutations of the template DNA (i.e. the coding region) or errors during transcription (the process of transcribing double stranded DNA into single stranded RNA, called messenger RNA or simply mRNA) can activate a cryptic splice site in which part of the transcript is usually not spliced. This is important to note, as a gene has introns and exons. Introns are that parts that do not contain coding sequences,

---

[36] Bioreactor:
An apparatus in which a biological reaction or process is carried out, especially on an industrial scale, according to Oxford Online Dictionary.

[37] http://www.it-asso.com/gxp/eudralex_v21/contents/vol-4/pdfs-en/anx02en200408.pdf

which are usually cut out unless an event triggering splice variants occurs. Exons contain the coding sequence of a gene. The result of the transcription process is the single stranded mRNA, which does not contain the introns[38]. Via the process of translation[39], the mRNA which migrated into the cytosol[40] will give rise to an enchained amino acid sequence.

Usually, N-terminal signalling sequences are responsible for the insertion of the elongating protein chain (based on amino acids) in the endoplasmatic reticulum[41]. The signal peptide[42] sequences are subsequently removed from the endoplasmic reticulum. This sequence is successively prolonged until the full protein sequence is formed; simultaneously the folding process of the protein occurs, which may be assisted by chaperones. Simplifying, the hydrogen bonding between different groups of amino acids of the protein sequence causes folding; in addition, there is the other basic driving factor of protein folding, hydrophobic interactions between hydrophobic amino acid groups of the protein sequence.

The folding process gives the protein its three dimensional structure. Complex proteins like antibodies, for example, are composed form a hetero-dimer of a light and a heavy chain. Thus four subparts are joined via disulfide[43] bonds to form a complex molecule.

The formation of disulfide bonds is a posttranslational modification and leads to structural changes in one protein chain or the joining and inter-connecting of two or more distinguished protein chains. External factors such as temperature, electric or magnetic fields and special limitations have an impact on protein folding and can cause mis-folding of the protein.

The assembly of proteins, either in the ER or the Golgi apparatus[44], creates a series of refinements, called posttranslational modification, which can include folding, cutting, joining and modifying of proteins to form the required variant (also called isoform) of the final protein.

---

[38] Intron:
A segment of a DNA or RNA molecule which does not code for proteins and interrupts the sequence of genes. Oxford Online Dictionary.

[39] Translation:
The process by which a sequence of nucleotide triplets in a messenger RNA molecule gives rise to a specific sequence of amino acids during synthesis of a polypeptide or protein. Oxford Online Dictionary.

[40] Cytosol:
The aqueous component of the cytoplasm of a cell, within which various organelles and particles are suspended, according to Oxford Online Dictionary.

[41] Endoplasmic reticulum (ER):
A continuous membrane system that forms a series of flattened sacs within the cytoplasm of a eukaryotic cell and is important in the biosynthesis, processing and transport of proteins and lipids. Britannica Academic Edition.

[42] Peptide:
A compound consisting of two or more amino acids linked in a chain, the carboxyl group of each acid being joined to the amino group of the next by a bond of the type –OC-NH-. Oxford Online Dictionary.

[43] Disulphide; A sulphide containing two atoms of sulphur in its molecule or empirical formula. Oxford Online Dictionary.

[44] Golgi apparatus:
Membrane-bound organelle of eukaryotic cells that is made up of a series of flattened , stacked pouches called cisterne. Britannica Academic Edition.

Posttranslational modification of proteins can give the protein a manifold means to exercise its biological functions. Usually, biologically active groups are added to the protein such as acetate, phosphate[45], various lipids[46]and glycans.

Glycosylation is a post translational modification resulting from the addition of a carbohydrate (glycosyl) group to exposed amino acids, such as asparagine, cysteine, hydroxylysine, serine, threonine, tyrosine or ortryptophan.

For mAbs, the glycosylation is of major importance for its biological actions. The amount and complexity of glycosylation differs, which results in different isoforms for the same basic protein skeleton. This causes considerable weight differences and possible disparity in biological activity patterns. For example, the weight of mAbs (ca. 145 kDA, IgG) can vary by 1000 Dalton due to heterogeneity (40).

The effects of this glycosylation on Abs are manifold. The circulation rate of an Abs in the vascular system is largely influenced by its glycosylation; an increase of the degree of glycosylation (sial-groups) determines the decrease in the clearance rate. This results in better bioavailability[47]. The glycosylation also affects the effector function aspects of Abs. For example, antibody-dependent cellular cytotoxicity (ADCC) was shown to vary inversely with respect to the Ab's glycosylation level.

> *"Rituximab was produced from Chinese hamster ovary (CHO) cells with a relatively high level of glycosylation and was found to be several-fold less cytotoxic*[48] *in vitro than another anti-CD20 mAb (KM3065), which was synthesized with fewer glycan residues from a rat cell line".* (41)

2.2.2 Variation in the biotechnology processes

Modern biotechnological processes have generated a wide range of novel products, including antibiotics, recombinant proteins, vaccines and monoclonal antibodies. Pharmaceutical biotechnology typically consists of cell banking, upstream (production) and downstream (separation/purification) operations.

---

[45] Phosphate: A salt or ester of phosphoric acid, containing PO43− or a related anion or a group such as ▯OPO(OH)2. Oxford Online Dictionary.

[46] Lipid: Any of a class of organic compounds that are fatty acids or their derivatives and are insoluble in water but soluble in organic solvents. They include many natural oils, waxes, and steroids. Oxford Online Dictionary.

[47] Bioavailability: The proportion of a drug or other substance which enters the circulation when introduced into the body and so is able to have an active effect., according toOxford Online Dictionary.

[48] Cytotoxic: Toxic to living cells, according to Oxford Online Dictionary.

2.2.2.1 Cell banking issues

Once the target gene is isolated and inserted into a suitable host cell for expression, a master cell bank is established. This cell bank will serve as the genetic source for all subsequent manufactured target proteins. However, technical issues are associated with creating stable cell lines that secrete the desired (i.e. previously inserted by recombinant DNA technology) target protein. (42)

Variation occurs between the different cells of the master cell bank due to genetic heterogeneity in the original cell line from which the master cell bank was derived, prior to insertion of the target gene.

> *"In addition to genetic heterogeneity, a significant fraction of total variation may arise from phenotypic differences between cells in each pure clone making up a pool. This, in turn, appears to result from random expression fluctuations in individual cells over time, as elegantly demonstrated in the landmark study of Sigal et al".* (43)

A biosimilar developer will not be able to reproduce the reference biologic 100%, including all is heterogeneity, as different cell lines will likely be used to manufacture the target protein.

> *"Particularly, transfection of the host cell represents a unique event which cannot be identically replicated – if done twice, the result would be a manufacturing cell line with different properties. The conditions used for cell fermentation will depend on the properties of the master cell bank and therefore cannot be identical for a second manufacturing process. In addition, the details of upstream and downstream manufacturing and the methods and criteria chosen for in-process analytics are not in the public domain and are therefore not known to a second, independent manufacturer".* (44)

2.2.2.2 Upstream issues

Upstream, the cells are thawed and multiplied in a series of volume increasing pre-cultures, after that they are transferred into the large scale production bioreactor. The final protein product is influenced by physico-chemical factors, such temperature, pH, osmolarity[49], CO_2, O_2 saturation and oscillation rate of the propeller, operation related factors, such as the type and mode of operation of the bioreactor (including the time in which the biopharmaceutical process is conducted in the bioreactor) and chemical factors derived from the cell culture media such as growth factors, carbohydrates, essential metals and vitamins, all affect the finial protein product.

At this stage, those external factors influence the cells during their division and protein expression phases. As mentioned in section: 2.2.1 Heterogeneity of biologics (proteins only),

---

[49] Osmolarity:
The concentration of a solution expressed as the total number of solute particles per liter, according to Oxford Online Dictionary.

this influences the protein formation and its subsequent post translational modifications. The importance of this was already discussed in the example of mAbs.

> *"Cell stress can lead to aberrant glycosylation, protease activity, protein truncation, disulfide shuffling, and the release of undesirable by-products and cellular wastes".* (45)

2.2.2.3 Downstream issues

The aim of downstream processing is to bring the target protein through progressive refinement steps to the required purity and concentration level. The harvesting process is initiated, depending on the type of bioreactor and the mode of operation, when certain key parameters are reached. The cells and supernatant (i.e. the remaining media) are separated and either the media or cell part is discarded, depending on whether the target protein is extracellularly or intracellularly expressed.

The target protein is further purified by adsorption, ultra-filtration and precipitation in order to reduce the volume and to increase the concentration of the target protein. Further purification of the bio-product can be achieved through a series of separation techniques that rely on the unique physico-chemical properties of the target protein, for example, affinity-, size exclusion- or HPLC-chromatography. These chromatography steps may be altered by ultra-filtration steps to exchange buffer and/or to increase the concentration of the target protein. These separation techniques aim to separate process related proteins from the host cell (different form the target protein), host cell DNA and other impurities will be present.

However, the uniqueness of the separation cascade will most likely result in the creation of a unique product, from the point of view of the isoform composition. Any change in the purification process can alter the purity of the product. (46)

Additional variation occurs in the target protein during downstream processes such as oxidation and de-amination. Of equal concern is the formation of protein aggregates, which have a major impact on the bioavailability and immunogenicity of the medicinal product and degradation of the target protein. (41, 45)

For the production of the active substance into the medical product, there are usually excipient and additives added to the formulation. At that stage issues may occur, as the additives can impact the safety profile.

> *"Initial analyses of a biosimilar's active ingredient may indicate acceptable similarity to the reference molecule, but the final product formulation may reveal potential risk factors. Absorption spectra of different formulations of similar products can reveal differences between them and identify chemicals and impurities that may have leached into the formulation".* (40)

2.2.2.4 Storage and stability issues

Final processing ends with conserving and packaging the target protein in a stable form that is easily transportable and convenient. Crystallization through lyophilisation is commonly used. However, some bio-products may be frozen in their buffer, or are aseptically processed and sterilely filtered. Once reconstituted and made ready for use (i.e. finial medicinal product), the stability is an issue, and the medicinal product is affected by environmental factors, and must be stored and transported under optimal conditions.

> *"A biopharmaceutical from the same manufacturer may also degrade at different speed and at different conditions due to inadequate control of the production process. Decreased stability in comparison with the reference product may either be due to an unstable formulation or an active ingredient".* (40)

2.2.3 The biologics life cycle

Broadly outlined, the biologics life cycle follows the life cycle of any other medical product. In the past, the one main difference was that there were practically no follow on versions (biosimilars) available, as there are in the case of innovator chemical medical products produced by generic companies. The generic industry could not simply copy a biologic like it did for small drug using only pharmacopeia specification. The complexity of biologic does not allow such an approach.

The life cycle of a biologic starts with a R&D phase, were the entity is discovered, and the proof of principles and concepts is performed.

In the next stage, the product and the process are developed. Usually, the process is successively scaled up, which often involves the usage of different processes and equipment. During that phase of the process, understanding, a prerequisite for a successful process validation, is generated.

In parallel, the pre-clinical testing kicks off, this compromises of animal studies to toxicity studies, as well as studies of pharmacological effects through in vitro and in vivo laboratory animal testing. Genotoxicity testing is carried out as well as investigations on the absorption and metabolism rates.

Then, when the process is running at a pilot plant level with controls on parameters relevant for patient safety, such as virus inactivation steps, filtration steps and quality control tests, for example, for endotoxines; the phase I studies are initiated with healthy volunteers. The main aim is to prove the basic safety of the product. Potential immunogenic complications are assessed, and some metabolical testing is carried out.

During phase II, the clinical trials consist of controlled clinical studies intended to answer particular questions on the effectiveness of the biologic.

In phase III, the process should be at a commercial scale, which is ready to be validated. Usually, concurrent to phase III, the validation process is carried out. Phase III trials are intended to gather additional information about the effectiveness and safety, and to contribute data to evaluate the benefit-risk of the biologic.

The time frame to obtain a MA can cover many years, refer also to section: 1.2.1.1 Outline of a biologics life cycle.

For example, the discovery and development process for Avastin (bevacizumab) from Roche (a mAb for cancer indication) took many years, as follows:

- After isolation of the vascular endothelial growth factor (VEGF) an anti-VEGF antibody was published in 1993.
- In 1996 the mAb was humanized (modified to be less immonugenetic).
- In 1997 an investigational new drug application for this antibody was submitted to the FDA, and a phase I trial for bevacizumab began in 1997.
- Phase II started in 1998.
- Phase III started in 2000.
- In February 2004, the FDA approved bevacizumab as the first anti-angiogenic drug for treating cancer, 15 years after the first isolation of VEGF. (47)

Upon successful approval by a regulatory authority, the commercial phase of a medicinal product begins. This is an area that is more relevant to the topic of this research. During its lifecycle, the biologic may change its quality profile. In fact one may distinguish between:

1) Normal lot to lot variability, due to the heterogeneity of the biologic itself and minor process variation (fluctuation). This normal variability is covered through the MA authorisation.

2) Factor dependent variation that leads to a more significant change of a biotechnological process. This can be due to inspection outcome, changes in equipment, shortages of a certain raw material, closedown of a facility and process transfer to another facility or process improvements, etc., the process might change.

It is well known that these changes need to undergo regulatory approval and require a comparability exercise in the case of biologics. The extent of the comparability exercise depends on the scope of the change and the complexity of the biologic itself. It has been argued that, *"these changes may lead to differences just as they occur in independently manufactured biosimilar products"*. (44)

For example (21), it was reported that innovator companies may also fail comparability exercises required by the FDA in the case of a product transfer.

"The FDA ruled that the Genzyme's Myozyme (alglucosidase alfa) treatment for Pompe disease produced at its facility in Allston, MA, was actually a different product than its Myozyme product already registered and produced at its facility in Framingham, MA. The agency determined that because of "differences in the biological signature of the active molecule" and concerns about how the protein is glycosylated, the same drug produced in two facilities within the same company were in fact not "similar"". (21)

2.2.4 Discussion

As outlined above, heterogeneity will result from complex biological processes, but is also interlinked to the environment, which stands for certain process conditions, for example. Heterogeneity can result from the manufacturing process and/or storage of the product.

"The pattern and variability of protein microheterogeneity will depend on the way how the protein is manufactured. Not all of these variations will have an impact on the clinical safety and efficacy profile, but since only a limited number of attributes can be assessed analytically, one cannot assume a priori that proteins which are produced using different processes have identical properties". (44)

Thus, it is required to characterize all isoforms of a biologic, and degradation products should be well characterized. If the process is altered or degradation products change, the quality profile should be compared to the original pre-change product. The example, the glycosylation level of mAbs has a major influence on the receptor binding (Fc-Receptor binding), protein protein interaction (Ab-Antigen interaction) and pharmacokinetics of the protein substances (circulation time defining its half-life and therefore its bioavailability). By that, mAbs immune effector functions, such as antibody-dependent cellular cytotoxicity (ADDC), which is a major mechanism of action, are influenced, refer also to section: 4.3 Results for main research questions 1.0 and directly associated research questions 1.1 and 1.2 that discus the impact of the dynamic to the quality profile and section: 4.4 Discussion when discussing the impact of those minor differences, which occur during the life-cycle.

Variation, as described above, is mainly related to the process, and possibly even more so, to the downstream process during which the purification occurs. For example, the employed cascade to purify the protein of interest may be unique and difficult to replicate, especially when taking the raw materials, which are required for that purification cascade, into account.

However, prior to discussing the impact of heterogeneity and variation one needs to become clear on how this relates to the research questions, and this requires a discussion of how this heterogeneity and variation can be viewed, and the resulting issues, when interpreting regulatory guidance documents. The summary of analyzable heterogeneity and variation forms what is called a quality profile.

However, there are undoubtedly two aspects to consider:

1. **The quality profile of a batch: the quality profile of a batch is fixed and analyzable.**
2. **The quality profile of a biologic: the quality profile of a biologic is dynamic in time as the summary of the quality profiles of all batches, and is therefore not fully analyzable.**

Refer to section: 4.3.2 Results from the questions and comments submitted to EMA, when discussing those aspects.

> *"It should be emphasized that batch-to-batch variations are unavoidable even with well-controlled, consistent protein manufacturing procedures and are acceptable within the limits defined by the specifications, but usually they are much smaller than differences between products made by totally different manufacturing processes".* (44)

This implies that there is not only within one batch level presence of variation and heterogeneity (isoforms and impurities for example); an even more complex situation raises where the degree and possible nature of heterogeneity and variation between batches varies (different ratio and identity of isoforms and different ratio and identity of impurities).

This results in variable quality profiles, when considered on a life cycle bases, with possibly different ratios and glyco-forms of the protein of interest (isoforms). As outlined above, the impact of even small structural differences can be clinically relevant. For example, differences in the glycosylation can affect the efficacy of a biologic, by influencing the mode/intensity of the action between molecules and their circulation time, as outlined above.

It is important for the conceptual understanding of the research question to understand this notion.
One result of heterogeneity and variation is the influence on the safety profile. Biologics as bio-molecules have the inherent risk of causing immune responses, either against the protein based active substance (including all its isoforms) or against an impurity or added excipient.

2.3 Screening the above presented literature related to current biosimilar regulation with regard to the research questions

The regulatory literature was selected based on Figure: 3 in section: 2.1.2 Regulatory guidance literature. All overarching and general guidance literature was reviewed with regard to text passages relevant for the research questions of this study:

1.0 What are the implications of heterogeneity and variation through the life cycle of the biosimilar and the reference biologic, from a European perspective?

1.1 Taking into account the new amendment of EMA/CHMP/BWP/617111/2010, with emphasis on the life-cycle of the Biosimilar, does the Biosimilar need to achieve a quality profile which falls within the quality profile of the reference biologic? Under which circumstances could this approach be deviated from?

1.2 What is proposed if the reference biologic changes its quality profile during the biosimilar development program?

2.0 What should be the scope of trials for biosimilars?

2.1 How extensive do biosimilar trials need to be in order to show comparability?

2.2 Under what circumstances does one need to test multiple lots of the biosimilar vs. the reference medicinal product in trials?

2.3 Meta Analysis, maybe a sign for extensive trials, as such reflecting potential issues that show comparability at a first glance.

3.0 Why is extrapolation of indications for biosimilar controversial?

This also included concept papers and draft guidance. Further, one product specific guide was reviewed, EMA/CHMP/BMWP/403543/2010 Guideline on similar biological medicinal products containing monoclonal antibodies – non-clinical and clinical issues (48). The reason for this is that the product specific guide was for the latest adopted biosimilar guide from EMA. Questions and comments were submitted to the EMA for this guide as part of this study and the contents of the guide itself are very relevant for the research questions.

Other product specific biosimilar guides were not included in this literature review, since the burden to review them was judged to be too extensive.

Directly after the selected text passage, comments and conclusions from the author were made. The headlines of the selected sections were then grouped and ordered in a table. Subsequently the findings were briefly summarised and discussed as part of Chapter 4.0: Main research question for research question 1, Chapter 5.0: What should be the scope of trials? for research question 2 and Chapter 6.0: Why is extrapolation of indications for biosimilar controversial? for research question 3.

2.3.1 Guideline on similar biological medicinal products CHMP/437/04 (49)

General comment:
In the context of the surrounding biosimilar documentation, this guidance document can be considered as lacking clarity. Most aspects deal with the quality profile and are of utmost importance for a successful biosimilar development program.

Emphasis is also put on characterisation whereas in other documents it is fully acknowledged that biologics can rarely be fully characterised, given the complexity and structural differences between large biologics such as proteins or even vaccines, which are not mentioned as part of recent EMA legislation.

Same reference material:
There is a section requiring that during the whole comparability exercise, the "*same reference product*" should be used for all three parts of the dossier, i.e. quality, safety and efficacy.

This is not possible as a development program can take years to be performed and the quality profile of the reference biologic can change at any stage without any influence from the biosimilar developer.

Conclusion/Comment:
It shows that the reference will change in the foreseeable future, but maybe those changes are considered to lead to a similar clinical outcome. If so, then, depending on the complexity of the biologic in question, that approach is likely to fail.

There is no guarantee that this approach is feasible for the future, and likely clinical comparability for more complex biologics will fail or is going to require multiple and repeat trials as data might not be fully conclusive.

Heterogeneity and variation:
As biologics are variable and the EMA acknowledges this variability, EMA argues that the success of a biosimilar program is based on the ability to characterise and reference the biosimilar; and therefore, show the similar nature of the products concerned. This was partially outlined in the sentence below:

> *"In principle, the concept of a "similar biological medicinal product" is applicable to any biological*
> *medicinal product. However, in practice, the success of such a development approach will depend on the ability to characterise the product and therefore to demonstrate the similar nature of the concerned products".*

This is also confirmed in the recent draft for the quality guide: 2.3.6 Guideline on similar biological medicinal products containing biotechnology-derived proteins as active substance:

Quality issues (revision 1) Draft EMA/CHMP/BWP/247713/2012 (54), for example. However, the EMA acknowledges in the same guide the difficulties encountered to characterise a biologic in 2004 by mentioning:

> *"Biological medicinal products are usually more difficult to characterise than chemically derived*
> *medicinal products. In addition, there is a spectrum of molecular complexity among the various*
> *products (recombinant DNA, blood or plasma-derived, immunologicals, gene and cell-therapy, etc.)".*

Conclusion/Comment:
This represents a good example of regulatory requirements as they are laid down in a formal way and subsequently when faced with reality/nature, adaptation and sensible flexibility is required by the regulator. The extent of similarity is, and will remain, difficult to be described.

Changes (issue of dynamic of the quality profile) during the life cycle:
With regard to the life cycle EMA also acknowledges early that, "changes can impact on the *three-dimensional structure, the amount of acido-basic variants or post-translational modifications such as the glycosylation profile*".

Conclusion/Comment:
The term "changes" reflects the words 'heterogeneity' and 'variation' employed in this study, based on my understanding when reading the context. Interestingly, EMA employed the term 'change', which, in the regulatory context, comes from a situation where a M.A. is already issued for a medicinal product and something potentially impacting quality, safety or efficacy changes (usually a process change). This may reflect the EMA's limited experience with biosimilars at the time, and the similarity to the guidance related to the comparability exercise, refer to section: 2.1.2 Regulatory guidance literature.

> *"Those changes which may initially be considered to be 'minor' in the manufacturing process. Thus, the safety/efficacy profile of these products is highly dependent on the robustness and the monitoring of quality aspects".*

Highly purified products:
The EMA outlines further that:

> *"Due to the complexity of biological/biotechnology-derived products the generic approach is scientifically not appropriate for these products. The "similar biological medicinal products" approach, based on a comparability exercise, will then have to be followed.*
> *– Comparability exercises to demonstrate similarity are more likely to be applied to highly purified products, which can be thoroughly characterised (such as some biotechnology derived medicinal products)".*

Conclusion/Comment:

The term 'highly purified' needs to be set in the context of the research question as being very arguable and misleading. In later guidance, the EMA requires a thorough evaluation and discussion. In situations where the biosimilar is more or less pure refer to Guideline on similar biological medicinal products containing biotechnology-derived proteins as active substance: Quality issues (revision 1) Draft EMA/CHMP/BWP/247713/2012. Surprisingly, EMA's guide was only updated in 2013.

EMA even though puts emphasis on the analytical characterisation and mentions it as a prerequisite for highly purified protein, it is also recognised that the fact that each biologic has its particularities is due to normal heterogeneity and variation. Therefore, it is written in the EMA guide that:

Highly purified products:

> *"It should be recognised that, by definition, similar biological medicinal products are not generic medicinal products, since it could be expected that there may be subtle differences between similar biological medicinal products from different manufacturers or compared with reference products, which may not be fully apparent until greater experience in their use has been established".*

Conclusion/Comment:

In the situation when dealing with references biologics, which themselves are not highly pure, the notion of 'high purity' and biologics is in this context is that only one isoform or sub type of molecule is being dealt with, is contradicted by other the EMA documents Guideline on comparability of medicinal products containing biotechnology derived proteins as active substance: Quality issues EMEA/CPMP/BWP/3207/00 and Concept paper on the revision of the guideline on similar biological medicinal product EMA/CHMP/BMWP/572643/2011.

Comparability exercise:

The EMA outlines in this overarching guide again the import factors for the success of the biosimilar approval.

1. Analytical procedures
2. The manufacturing processes
3. Clinical and regulatory experiences

Conclusion/Comment:

None

Clinical aspects:

The clinical aspect is then further outlined:

"The requirements to demonstrate safety and efficacy of similar biological medicinal products have to comply with the data requirements laid down in Annex I to Directive 2001/83/EC".

Conclusion/Comment:

These aspects should be understood in conjunction with other documents, refer to Guideline on comparability of medicinal products containing biotechnology derived proteins as active substance: Quality issues EMEA/CPMP/BWP/3207/00 and Guideline on similar biological medicinal products containing biotechnology-derived proteins as active substance: Non-clinical and clinical issues EMEA/CHMP/BMWP/42832/2005. In other words, EMA expects the safety and efficacy to be comparable within margins of tolerance, but does not expect proof of safety or efficacy per se, as this was already proven by the reference biologic itself.

Other aspects:

It was also announced in this guide that EMA would issue product-class specific guidance.

Table: 3

Summary of Guideline on similar biological medicinal products CHMP/437/04 (49) with regard to the research question of this study.

Research question	**Input from the guide**
What are the implications of heterogeneity and variation through the life cycle of the biosimilar and the reference biologic?	Same reference material throughout the comparability exercise. Heterogeneity and variation Changes (issue of dynamic of the quality profile) during the life cycle Highly purified products Comparability exercise
Does the Biosimilar need to achieve a quality profile which falls within the quality profile of the reference biologic? Under which circumstances could be this approach be deviated from?	
What is proposed if the reference biologic changes its quality profile during the biosimilar development program?	
What should be the scope of trials?	Clinical aspects
Extrapolation	None

2.3.2 Guideline on comparability of medicinal products containing biotechnology derived proteins as active substance: Quality issues EMEA/CPMP/BWP/3207/00 (50)

The guide on comparability, "Guideline on comparability of medicinal products containing biotechnology derived proteins as active substance", came into effect in 2003.

Highly purified proteins:
The reference is to highly pure molecules (as in section: 2.3.1 Guideline on similar biological medicinal products CHMP/437/04) and the consistency that forms between batches.

Conclusion/Comment:
Both are achievable only when giving them biologically meaningful margins. However they remain ideals, barely achieved.

The guide also mentions that the main tasks will be to establish to what extend the analytical methods used are able to detect any slight modification possibly introduced by the change.

This is an issue for biosimilars too. Mainly because after the initial approval of the reference biologic, many years will elapse and the analytical methods will be more evolved, thus always detecting differences within the quality profile of biologics. In cases where no differences are detected, the method's sensitivity should be questioned.

On the other hand, those differences in the quality profile have an impact on the clinical outcome. This answer remains open and is further analysed using the questionnaire, refer also to section: 4.3.3 Results from the survey research method. (50)

Comparability in relation to isoform ration/process changes for an already approved biologic:
The "Guideline on comparability of medicinal products containing biotechnology derived proteins as active substance" mentions there is a case to consider conducting comparability studies when seeking a marketing authorization for a biotechnology derived product claimed to be similar to one already authorised.

Conclusion/Comment:
This aspect needs to be considered in context, if, for example, the production process of the innovator biologic changes, for example, the production site changes. In principle, biosimilar production could be regarded as a similar case. However, the change it represents to the production process is still reasonably considered as being of a lower impact compared to the change the biosimilar itself represents, compared to the reference biologic.

However, from a life-cycle perspective this guide outlines the requirements for changes in the production process of an approved biologic. The comparability exercise principles apply to both the biosimilar and the reference biologic.

Five stages were mentioned for conducting the comparability exercise:

1) Characterisation studies
2) Validation manufacturing process
3) Release data
4) Stability data
5) Clinical and preclinical studies

The definition of a comparability exercise is outlined in the guide as the follows: *"Comparability is the exercise that will demonstrate that two products have similar profile in terms of quality, safety and efficacy"*. It has already been outlined that, possibly next to quality data, clinical data might be required.

Conclusion/Comment:
As outlined in section: 2.1.2 Regulatory guidance literature, the complexity of the molecular entity should be considered as a major criterion when discussing the comparability.

The guide mentions one particular train of thought worth mentioning as directly relevant to the study:

> *"In many cases due to inherent variability of the biotechnological process, the end-product consists of a complex mixture of molecules **(product-related substances).** This heterogeneity, which is taken into account when assessing the in-vivo behaviour of the product should be characterised to assure batch to batch consistency"*.

Conclusion/Comment:
This can be interpreted as the ratio of the isoform that should be consistent over batches, refer to section: 4.1 Introduction. During the life-cycle of a biologic, heterogeneity and variation always occur; however, the overarching biosimilar guide (49) does as well as this guide does in not satisfactorily acknowledging the fact and does not reflect the biological realities. (50)

Clinical trials and comparability exercise depend on:
The 3rd chapter of the guide mentions that the process is different and comparing to public standards is not sufficient. Thus, an extensive comparability exercise is required. The extent of pre-clinical and/or clinical testing need to be considered with regard to the complexity of the biopharmaceutical itself, as well being based on possible differences in the reference biologic itself; from the point of view of this study, those differences require a thorough investigation.

The following factors should be taken into account for a comparability exercise:
1) The complexity of the molecules
2) The type of changes in the process
3) Their impact on quality, safety and efficacy

Conclusion/Comment:

It is acknowledges that each individual situation requires a step-by-step approach. However, the way in which the guidelines were formulated does not apply to all the potential issues. This statement could be judged as a backdoor for discussion with EMA in cases where the limitations of the comparability exercise were reached, which cannot be excluded. (50)

Table: 4
Summary of Guideline on comparability of medicinal products containing biotechnology derived proteins as active substance: Quality issues EMEA/CPMP/BWP/3207/00 (50) with regard to the research question of this study.

Research question	Input from the guide
What are the implications of the heterogeneity and variation through the life cycle of the biosimilar and the reference biologic? Does the biosimilar need to achieve a quality profile that falls within the quality profile of the reference biologic? Under which circumstances could the approach be deviated from?	**Highly purified proteins:** **Comparability in relation to Isoform ration/process changes for an already approved biologic:** *"In many cases due to inherent variability of the biotechnological process, the end-product consists of a complex mixture of molecules **(product-related substances).** This heterogeneity, which is taken into account when assessing the in-vivo behaviour of the product should be characterised to assure batch to batch consistency"* This can be interpreted that the ratio of isoform should be consistent over batches, refer to research question 1.0.
What is proposed if the reference biologic changes its quality profile during the biosimilar development program?	None
What should be the scope of trials?	**Clinical trials and comparability exercise depend on this.** Thus, an extensive comparability exercise is required. The extent of pre-clinical and/or clinical testing should be a function of the complexity of the biopharmaceutical itself, as well as based on possible differences to the reference biologic itself.
Extrapolation	None

2.3.3 Guideline on comparability of medicinal products containing biotechnology-derived proteins as active substance: Non-clinical and clinical issues EMEA/CPMP/3097/02/Final (51)

Immunogenecity:
There is also a guide for the comparability exercise with regard to clinical and non-clinical issues available. Biosimilars were mentioned as well. The main concern was on safety and immunogenicity with regard to biosimilars.

However, there were no requirements outlined for efficacy. Input or discussion on clinical trial requirements impacting clinical trial size were not detected. Thus, this guide remains of a more abstract nature with regard to the research questions of this study.

Conclusion/Comment:
None

Table: 5
Summary of Guideline on comparability of medicinal products containing biotechnology-derived proteins as active substance: Non-clinical and clinical issues EMEA/CPMP/3097/02/Final (51) **with regard to the research question of this study.**

Research question	Input from the guide
What are the implications of the heterogeneity and variation through the life cycle of the biosimilar and the reference biologic?	None
Does the Biosimilar need to achieve a quality profile which falls within the quality profile of the reference biologic? Under which circumstances could this approach be deviated from?	
What is proposed if the reference biologic changes its quality profile during the biosimilar development program?	None
What should be the scope of trials?	Immunogenicity
Extrapolation	None

2.3.4 Concept paper on the revision of the guideline on similar biological medicinal product EMA/CHMP/BMWP/572643/2011 (52)

Highly purified proteins:

Compared to the current guide (2.3.1 Guideline on similar biological medicinal products CHMP/437/04 (49)) on that topic there were some significant changes made. For example the following section:

> *"Comparability exercises to demonstrate similarity are more likely to be applied to highly purified products, which can be thoroughly characterised (such as some biotechnology derived medicinal products). The 'similar biological medicinal product' approach is more difficult to apply to other types of biological medicinal products, which by their nature are more difficult to characterise, such as biological substances arising from extraction from biological sources and/or those for which little clinical and regulatory experience has been gained. Whether a medicinal product would be acceptable using the 'similar biological medicinal product' approach depends on the state of the art of analytical procedures, the manufacturing processes employed, as well as clinical and regulatory experiences".* (49)

The above is supposed to be have been amended and the following concept paper is available (52). In its problem statement, CHMP mentions that the way in which biosimilars are approved should be reflected in the guide. That indicates there is a discrepancy between current practice and guidance requirements, referring back to the case by case approach backdoor mentioned already in section: 2.3.3 Guideline on comparability of medicinal products containing biotechnology-derived proteins as active substance: Non-clinical and clinical issues EMEA/CPMP/3097/02/Final (51).

Conclusion/Comment:

Mention of highly purified proteins implies certain homogeneity of the active substance and its impurities.

Refinements for the biosimilarity approach:

> *"The principles of biosimilarity need to be better explained. Terminology issues such as "biosimilar" or "similar biological medicinal product", and often the term "biosimilar" has been used in an inappropriate way. Discuss the feasibility to follow the generic legal basis for some biological products. Some more specific aspects require re-discussion and potentially a refinement".*

Conclusion/Comment:

To recapitulate, the current biosimilars approach follows a train of thoughts that:

Clinical benefit has already been established by the reference medicinal product, and that the aim of a biosimilar development program is to establish similarity to the reference product, not clinical benefit. Those aspects are significant for all research questions, as it has an impact on the requirement to establish similarity and the strategies employed doing so using a comparability exercise.

Clearly, the requirement not only to understand the biosimilar but also the reference biologic itself in an extensive comparability exercise is exceeding the requirements a biologic developer would face when developing an innovator drug. However, the scope of trials is influenced by the intended purpose of the trial. That is common sense.

The scope of the trials is fundamentally different, as they are not required to show safety or efficacy, but similarity, thus the trial design is different and possibly smaller for a biosimilar when compared to an innovator biologic.

In this context, below are the EMA outlines:

> *"As regards more specific aspects, there are some items in the current text that require refinement, e.g.EMA/CHMP/BMWP/572643/2011 3/4*
> *A discussion of equivalence of efficacy and safety aspects, should this be necessary and not be covered by the revision of the general non-clinical and clinical guideline".*

However, the EMA does not provide information on more concrete issues with regards to non-clinical and clinical trial topics. The differences in PD or PK studies, due to heterogeneity and variation of biologics, were not discussed. Maybe because they do not occur, or because they have always occurred also for biologics M.A., and therefore, the EMA does not feel there is the need to discuss this further.

However as those tests have to be considered in a head-to-head comparison as part of a comparability exercise, the difference will show up and the potential impact remains unknown, as the safety and efficacy data for the reference biologic is limited by the given quality profile of the biologic.

Differences in the quality profile could lead to confounded different clinical issues. The term confounded has to be used, as when viewed on subject matter basis the experimental conditions for PK and PD studies are barely the same in a comparability exercise setting.

However, from a rationale basis, the differences in the quality profile must lead to differences in the PK and PD. The molecular structure influences those parameters heavily.

Table: 6
Summary of Concept paper on the revision of the guideline on similar biological medicinal product EMA/CHMP/BMWP/572643/2011 **(52) with regard to the research question of this study.**

Research question	**Input from the guide**
What are the implications of heterogeneity and variation through the life cycle of the biosimilar and the reference biologic?	Highly purified products Refinements for the biosimilarity approach
Does the Biosimilar need to achieve a quality profile which falls within the quality profile of the reference biologic? Under which circumstances could this approach be deviated from?	
What is proposed if the reference biologic changes its quality profile during the biosimilar development program?	
What should be the scope of trials?	
Extrapolation	None

2.3.5 Guideline on similar biological medicinal products containing biotechnology-derived proteins as active substance: Quality issues EMEA/CHMP/BWP/49348/2005 (53)

Heterogeneity and variation

"In addition, each medicinal product is defined by the molecular composition of the active substance resulting from its process, which may introduce its own process related impurities.
Consequently, the similar biological medicinal product is defined by the following two sets of
characteristics:
1) related to the characteristics of the molecule (including product related substances/impurities),
2) related to its process
(which may affect molecular characteristics and includes process related impurities)".

Conclusion/Comment:

Product related substance impurities could be interpreted as being host cell DNA, for example. However, it is not clear when talking about impurities what is actually meant. There is no line between the heterogeneity (normal biological quality profile), variation (introduced by the process, and according to annex II confounded with the product) and impurities (usually associated with the purification process itself).

Quality attribute comparison:

> *"A stepwise approach should be undertaken to justify any differences in the quality attributes of the similar biological medicinal product versus the reference medicinal product in order to make a satisfactory justification of the potential implications with regard to the safety and efficacy of the product".*

Conclusion/Comment:

Then there is a provision in the case where the quality attributes are not identical, which is complementary to the importance of the quality profile, which is highlighted as follows:

> *"It is not expected that the quality attributes in the similar biological and reference medicinal products will be identical. For example, minor structural differences in the active substance, such as variability in post-translational modifications may be acceptable, however, must be justified. Likewise, differences between the impurity profiles of the similar biological medicinal product and the reference medicinal product should be justified and would be considered on a case-by-case basis, and supported by the comparability exercise for quality attributes in relation to safety and efficacy.*

Same reference material:

> *"The same reference medicinal product must be used for all three parts of the dossier (i.e. Quality, Safety and Efficacy)".*

Conclusion/Comment:

The extent of sameness of the reference material remains unknown. Similar reference material can be used; however, the same, due to shelf-life limitations, is barely feasible. Further, from the conceptual point various lots should be compared to each other, including "several lots of the reference material set the boundaries of the biosimilar", refer to section: 2.3.6 Guideline on similar biological medicinal products containing biotechnology-derived proteins as active substance: Quality issues (revision 1) Draft EMA/CHMP/BWP/247713/2012 (54). Thus it is not clear again what is meant by the same reference material that is mentioned.

Heterogeneity of biologics

> *"Analytical methods must be suitable for the intended purpose, as the complexity of the molecule and its inherent heterogeneity required thorough analysis".*

Conclusion/Comment:
The EMA acknowledges that the molecules to be analysed during the comparability exercise, including the biosimilar itself, are heterogeneous. As the term 'inherent heterogeneity' is employed, it is clear that associated heterogeneity is meant and not impurities etc., as in other guidedance documentation, refer to section: 4.4 Discussion.

The biosimilar developer has the burden to demonstrate that the selected analytical methods used for the comparability exercise are capable of detecting the slightest differences in all aspects of the quality profile.

Product related substance vs. impurities:
Actually it is further specified in the section for physicochemical properties that:

> *"the evaluation of physicochemical parameters and the structural identification of product-related substances and impurities, including the determination of degradation by performing stress and accelerated stability studies. A physicochemical characterisation program should include a determination of the composition, physical properties, primary and higher order structures of the active substance of the similar biological medicinal product.*
> *An inherent degree of structural heterogeneity occurs in proteins due to the biosynthetic process, therefore, the similar biological medicinal product can contain a mixture of post-translationally modified forms. Appropriate efforts should be made to investigate and identify these forms"*.

This passage speaks clearly about isoforms, i.e. the same biosimilar active substance in all its variations (isoform are variations of a protein). Thus, clearly outlines the fact that EMA acknowledges the complexity and post transcriptional variability.

The purities and impurities, which in the view of this study were confused with the above aspects of heterogeneity and variation, should not be that strictly separated from aspects of purities or impurities:

> *"The purity and impurity profiles of the active substance and medicinal product should be assessed both qualitatively and quantitatively by a combination of analytical procedures for both reference and similar biological medicinal products".*

> *"The product-related substances and impurities in the similar biological medicinal product should be identified and compared to the reference product using state-of-the-art technologies. Additionally, information based on the analysis of samples stored under stress conditions, inducing selective degradation (e.g., oxidation, dimerisation) should be used for identification. Comparison of product related substances, and of product-related impurities should be based on specific degradation pathways and potential post-translational modifications of the individual proteins. Accelerated stability studies of the*

reference and of the similar biological medicinal product can be used to further define and compare stability profiles".

The EMA acknowledges that variation and process-related impurities, such as host cell proteins, host cell DNA and other impurities, are expected to be preened in the biosimilar as in any biologic and to differ from the quality profile of the reference biologic.

Further process-related impurities are required to be investigated and assessed by appropriate studies (which may include non-clinical and/or clinical studies).

Clinical scope:

"Therefore, differences in impurity profiles and significant differences in product related substances may have consequences with regard to the amount of non-clinical and clinical data which may be required in order to make satisfactory justification of the safety and efficacy of the similar biological medicinal product".

Conclusion/Comment:

The extent of heterogeneity and variation impacts the extent of the clinical trial requirements. It is required to show comparability of the similar biological medicinal product with the chosen reference product for both the finished medicinal product (of the reference biologic and the biosimilar) and the active substance (both active substances).

Table: 7
Summary of Guideline on similar biological medicinal products containing biotechnology-derived proteins as active substance: Quality issues EMEA/CHMP/BWP/49348/2005 **(53)** with regard to the research question of this study.

Research question	Input from the guide
What are the implications of heterogeneity and variation through the life cycle of the biosimilar and the reference biologic?	Heterogeneity and variation Quality attributes Same reference material Heterogeneity of biologics Product related substance vs. impurities
Does the Biosimilar need to achieve a quality profile which falls within the quality profile of the reference biologic? Under which circumstances could this approach be deviated from?	
What is proposed if the reference biologic changes its quality profile during the biosimilar development program?	None
What should be the scope of trials?	None
Extrapolation	None

2.3.6 Guideline on similar biological medicinal products containing biotechnology-derived proteins as active substance: Quality issues (revision 1) Draft EMA/CHMP/BWP/247713/2012 (54)

(Note: For this guide, questions and comments as part of this study were forwarded to the EMA.)

Quality profile:

> *"On the scientific knowledge gained from the reference medicinal product, provided that the active substance of the biosimilar has been demonstrated to be similar, in physicochemical and biological terms, to the active substance of the reference medicinal product".*

Conclusion/Comment:

This is directly relevant for research question 1 of this study (context: mimicking of the quality profile), as the scientific knowledge is another description of the quality profile, which is a more common and more easily understandable terminology.

It outlines that the quality profile of the reference biologic has to be similar to the quality profile of the reference biologic on an active substance basis. This is new compared to the current guide. It implies that the two biologics are required to be similar, but there is no mention of 'highly purified', which led to thinking that there was an issue with accepting the naturally occurring heterogeneity of the biosimilar and biologics in general.

No mentioning of highly purified biologics:

> *"The development and documentation for biosimilars should cover two distinct but complementary aspects:*
>
> *i) molecular characteristics and quality attributes (QA) of the target product profile should be comparable to the reference medicinal product;*
>
> *ii) performance and consistency of the manufacturing process of the biosimilar on its own".*

Conclusion/Comment:

There is no longer any mention of purity or highly purified protein etc. However specific ICH Q8 R2 language is used.

ICH Q8 R2 language:

> *"The quality target product profile (QTPP) of a biosimilar should be based on data collected on the chosen reference medicinal product, including publicly available information and data obtained from extensive characterisation of the reference medicinal product. The QTPP should be detailed at an early stage of development and forms the basis for the development of the biosimilar product and its manufacturing process. It is*

important to identify critical quality attributes that may impact the safety and efficacy of the product".

Conclusion/Comment:

The EMA utilises for this guide the words and aspects of ICH Q8 R2[50]. The QTPP is defined by EMA as the target quality profile for the biosimilars developer. Thus mean that the heterogeneity of the bio molecule and the variability of the innovator process are setting the boundaries for the biosimilar developer. Further, it is mentioned that critical quality attributes should be investigated and possible differences in the quality profiles must be justified.

However, if a design space [51]is applicable or even meaningful for a biosimilar, still remains unclear. Such a design space could represent a solution when the quality profile of the reference biologic changes during the biosimilar development program as the biosimilar developer could then adjust the process and thereby the quality profile to the reference biologic prior to entering clinical trials, refer also to the discussion in section: 4.4 Discussion.

Life cycle of the biosimilar:

"It is acknowledged that the biosimilar will have its own lifecycle".

Conclusion/Comment:

This means that a biosimilar once approved and found to be similar to the reference biologic will no longer be measured against the reference biologic.

Analysis of the reference biologic as the basis for the biosimilar development:

"Several different batches of the reference medicinal product should be used to provide a robust analysis and to generate a representative quality profile. The relative age of the different batches of reference medicinal product should also be considered when establishing the target quality profile".

Conclusion/Comment:

EMA extensively describes a development program and outlines that several batches should be used, which corresponds to information submitted to the EMA for mAbs in 2011, as part of

---

[50] ICH Q8 R2:
INTERNATIONAL CONFERENCE ON HARMONISATION OF TECHNICAL REQUIREMENTS FOR REGISTRATION OF PHARMACEUTICALS FOR HUMAN USE
ICH HARMONISED TRIPARTITE GUIDELINE PHARMACEUTICAL DEVELOPMENT Q8(R2)
Developed by the appropriate ICH Expert Working Group and has been subject to consultation by the regulatory parties, in accordance with the ICH Process. At Step 4 of the Process the final draft is recommended for adoption to the regulatory bodies of the European Union, Japan and USA

[51] Design space:
The multidimensional combination and interaction of input variables (e.g., material attributes) and process parameters that have been demonstrated to provide assurance of quality. Working within the design space is not considered as a change. Movement out of the design space is considered to be a change and would normally initiate a regulatory post approval change process. Design space is proposed by the applicant and is subject to regulatory assessment and approval. (ICH Q8 R2)

study. Also, the phrase 'quality profile' is used as the summary of characterization tests, thus this represents a change in EMA language.

Comparability:

> *An extensive comparability exercise will be required to demonstrate that the biosimilar has a highly similar quality profile when compared to the reference medicinal product. This should include comprehensive side-by-side analyses of the proposed biosimilar and reference medicinal product using sensitive and orthogonal methods to determine not only similarities but also potential differences in quality attributes. Any differences detected in the quality attributes will have to be appropriately justified with regard to their potential impact on safety and efficacy".*

Conclusion/Comment:

It is not clear if the quality profile itself is meant to mimic the heterogeneity and variation on a lot to lot basis. For example, the ratio of isoform may vary between each other, however, remains relatively constant in terms of identify of the isoform, refer also to the discussion in section: 4.4 Discussion.

Finished biosimilar to be highly similar:

> *"The applicant should demonstrate that the desired product and product-related substances present in the finished product of the biosimilar are highly similar to that of the reference medicinal product. Where quantitative differences are detected, such differences should be demonstrated to have no relevance for the clinical performance of the product. Qualitative differences of product-related substances and/or impurities) require a thorough justification, which may include non-clinical and/or clinical data, as appropriate. It is however preferable to rely on purification processes to remove impurities rather than to establish a preclinical testing program for their qualification".*

Conclusion/Comment:

This means that the purity and impurity profiles must be justified and shown to be similar; however, the extent of the similarity is not clear. Again does this relate to a particular isoform and if so to what extend are differences allowed?

The range of permitted variability:

> *"The target acceptance criteria used in the comparability exercise should be justified. Quantitative limits should be established, where possible. The relevance of these limits should be discussed, taking into account the number of reference medicinal product lots tested, the quality attribute investigated and the test method used. These limits should not be wider than the range of variability of the representative reference medicinal product batches, unless otherwise justified.*

Conclusion/Comment:

The quality profile of the reference biologic defines the acceptance criteria for the biosimilar, thus a sufficiently wide range is required to capture the possible, full range of the target profile. Refer also to section: 4.4 Discussion.

Life-cycle:

> *"It is acknowledged that the manufacturing process of the reference medicinal product may evolve*
> *Through its lifecycle, and May lead to detectable differences in some quality attributes. Such events could occur during the development of a biosimilar medicinal product and may result in a development according to a QTPP which is no longer fully representative of the reference medicinal product available on the market.*
> *The ranges identified before and after the observed shift in quality profile could normally be used to support the comparability exercise at the quality level, as either range is representative of the reference medicinal product. Quality attribute values which are outside the range(s) of variability measured in the different profiles of the reference medicinal product should be appropriately justified with regard to their potential impact on safety and efficacy.*
> *It should also be noted that there is no regulatory requirement for re-demonstration of biosimilarity once the Marketing Authorisation is granted."*

Conclusion/Comment:

The life cycle is set here in relation to the quality profile. This is the first mention of changes in the quality profile of the reference biologic during the development cycle of the biosimilar. However, depending on if and when this event may occur, the EMA does not provide further guidance on the issue. Refer also to section: 4.4 Discussion.

Biologic heterogeneity:

> *"The target amino acid sequence of the biosimilar should be confirmed and is expected to be the same as for the reference medicinal product. Any detected differences should be part of the micro-heterogeneous pattern of the reference medicinal product".*

Conclusion/Comment:

Heterogeneity can be translated within the protein based no biologic context as an isoform. Thus the sentence does not make sense as "a detected difference should be part of the exiting isoform pattern" and is forcibly part of the QTPP.

Mimicking the Q-profile:

> *"The age/shelf life of the reference medicinal product at the time of testing should be mentioned, and its potential effect on the quality profile should be discussed where appropriate. Comparison of relevant quality attributes, tested at selected time points and storage conditions (e.g. accelerated or stress conditions), could be used to further support*

the similarity of the degradation pathways of the reference medicinal product and of the biosimilar"

Conclusion/Comment:
There is the question again about mimicking the quality profile on a lot basis or mimicking the quality profile on a summary of lots basis, where the ratio and identity of the isoform may be different and wider. Again each lot of the reference biologic will vary as well as when compared to each other, thus there is the open question of acceptable biosimilarity limits.

Table: 8
Summary of Guideline on similar biological medicinal products containing biotechnology-derived proteins as active substance: Quality issues (revision 1) Draft EMA/CHMP/BWP/247713/2012 (54) **with regard to the research question of this study.**

Research question	**Input from the guide**
What are the implications of heterogeneity and variation through the life cycle of the biosimilar and the reference biologic?	Quality profile No mentioning of highly purified biologics ICH Q8 R2 language Life cycle of the biosimilar
Does the Biosimilar need to achieve a quality profile which falls within the quality profile of the reference biologic? Under which circumstances could this approach be deviated from?	Analysis of the reference biologic as the basis for the biosimilar development Finished biosimilar to be highly similar The range of permitted variability Life-cycle Biologic heterogeneity
What is proposed if the reference biologic changes its quality profile during the biosimilar development program?	
What should be the scope of trials?	None
Extrapolation	None

2.3.7 Guideline on similar biological medicinal products containing biotechnology-derived proteins as active substance: Non-clinical and clinical issues EMEA/CHMP/BMWP/42832/2005 (55)

Quality: Same reference material:

"The *<u>same reference medicinal</u>* *product should be used for all three parts of the dossier (i.e. quality. Safety and efficacy aspects)."*

Quality: Quality profile based on final production batches:

In any case, the company should justify the approach taken during the development of the product and might want to contact the EMEA before starting the development for scientific and regulatory advice. It is recommended to generate the required clinical data for the comparability study with the test product as produced with the final manufacturing process and therefore representing the quality profile of the batches to become commercialised".

Conclusion/Comment:
This clinical data shall be derived from the final quality profile.

Clinical comparability exercise:

"The design of comparative PK studies should not necessarily mimic that of the standard "clinical comparability" design (CHMP/EWP/QWP/1401/98), since similarity in terms of absorption/bioavailability is not the only parameter of interest. In fact, differences in elimination characteristics between products e.g. clearance and elimination half-life should be explored. The choice of the design for single dose studies, steady-state studies, or repeated determination of PK parameters should be justified by the applicant".

"The pharmacodynamic (PD) markers should be selected on the basis of their relevance to demonstrate therapeutic efficacy of the product. The pharmacodynamic effect of the test and the reference medicinal products should be compared in a population where the possible differences can best be observed. The design and duration of the studies must be justified. Combined PK / PD studies may provide useful information on the relationship between exposure and effect".

Conclusion/Comment:

- This is an indication that the clinical phase should be designed to detect differences in efficacy.
- PK studies should focus on detecting possible differences.

PK/PD studies instead of clinical trials:

"Normally comparative clinical trials are required for the demonstration of clinical comparability. In certain cases, however comparative PK/PD studies between the similar biological medicinal product and the reference medicinal product may be sufficient to demonstrate clinical comparability, provided that all the following conditions are met...."

Conclusion/Comment:

- Comparability can also be shown based on PD/PK studies (under certain circumstances), which leavers the scope of phase III comparability studies for efficacy.

Clinical trials:

"Usually comparative clinical trials will be necessary to demonstrate clinical comparability between the similar biological and the reference medicinal product. Clinical comparability margins should be pre-specified and justified, primarily on clinical grounds. As for all clinical comparability trial designs, assay sensitivity (see ICH topic E10) has to be ensured.

If a clinical comparability trial design is not feasible, other designs should be explored and their use discussed with the competent authorities. Data from pre-authorisation clinical studies normally are insufficient to identify all potential differences. Therefore, clinical safety of similar biological medicinal products must be monitored closely on an ongoing basis during the post-approval phase including continued benefit-risk assessment".

Conclusion/Comment:

- Differences due to insufficient similarity should be detected prior to the M.A., thus the EMA acknowledges the limitations of that concept and defines an increased monitoring interval to complement the similarity exercise.
- Clinical comparability trials are usually required, but this can be amended if not feasible.

Immunogenicity in safety trials:

"The periodicity and timing of sampling for testing of antibodies should be justified. In view of the unpredictability of the onset and incidence of immunogenicity, long term results of monitoring of antibodies at predetermined intervals will be required. In case of chronic administration, one-year follow up data will be required pre-licensing".
"On the scope of trial also antibody testing is required not only for Abs against the biosimilar itself, but also for process related impurities".

Conclusion/Comment:

- Long term (up to 1 year) sampling for antibodies against the biosimilar and impurities are required.

Extrapolation of indication:

"It is important to consider the risk of immunogenicity in different therapeutic indications separately".

Conclusion/Comment:

- This could relate to the issue that different patient groups with different underlying health issues could be treated with one biologic, but under different indications. Thus, a separate assessment might be required that the biosimilar is safe for each group of patients, depending on their respective conditions that are typical for their disease type.

Extrapolation of indication depends on also on safety aspects of subpopulation:

"Possible safety issues in different subpopulations should also be addressed".

Conclusion/Comment:

- Extrapolation of indication depends also on the safety aspects of the subpopulation.

Extrapolation of indication:

"In case the originally authorised medicinal product has more than one indication, the efficacy and safety of the medicinal product claimed to be similar has to be justified or, if necessary, demonstrated separately for each of the claimed indications.

In certain cases it may be possible to extrapolate therapeutic similarity shown in one indication to other indications of the reference medicinal product. Justification will depend on e.g., clinical experience, available literature data, whether or not the same mechanisms of action or the same receptor(s) are involved in all indications."

Conclusion/Comment:

- Extrapolation of indication remains controversial, especially when different isoform from complex bio-molecules, such as mAbs, have different indications. Even though the mechanism of action might be the same, PD and PK might be very different for different isoforms. As for one indication, the quality data is further substantiated using clinical trials, lack of clinical trials might result in a biosimilar product that has a different efficacy, as the ratio and identity of isoform compared to the reference is different.

Table: 9
Summary of Guideline on similar biological medicinal products containing biotechnology-derived proteins as active substance: Non-clinical and clinical issues EMEA/CHMP/BMWP/42832/2005 (55) with regard to the research question of this study.

Research question	Input from the guide
What are the implications of heterogeneity and variation through the life cycle of the biosimilar and the reference biologic?	None
Does the Biosimilar need to achieve a quality profile which falls within the quality profile of the reference biologic? Under which circumstances could this approach be deviated from?	
What is proposed if the reference biologic changes its quality profile during the biosimilar development program?	
What should be the scope of trials?	IMMUNOGENICITY in safety trials
Extrapolation	Extrapolation of indication Extrapolation of indication depends on also on safety aspects of subpopulation

2.3.8 Concept paper on the revision of the guideline on similar biological medicinal products containing biotechnology derived proteins as active substance: Non-clinical and clinical issues EMA/CHMP/BMWP/572828/2011 (56)

Quality of the biosimilar:

"Biosimilar products produced with modern technologies may result in a reduced immunogenicity as compared to the reference product, and the BMWP will discuss if these products would still qualify as biosimilar".

Conclusion/Comment:

- So does a bio better qualify as a biosimilar? That would be a breakaway form the biosimilarity concept, towards a concept of relying on clinical data only, which would be based on non-inferiority trials.

Scope: A new concept is available for the scope of clinical trials:

"With regard to the phase I studies the guideline recommends the selection of relevant pharmacodynamic markers, but such markers are not always available and/or cannot be shown to reflect efficacy. On the other hand, when relevant surrogate markers do exist, the current guideline includes the possibility of using them (e.g. absolute neutrophil [52]count to assess the effect of granulocyte-colony stimulating factor) as the primary end points in the pivotal phase III studies. This aspect could be expanded to further elaborate on the underlying principles, potentially including examples from other therapeutic areas e.g. oncology. The current guideline does not mention any recommendations on the patient population in the pivotal phase III studies".

Conclusion/Comment:

- Only phase I and phase III trials were mentioned. Phase II trials were not mentioned and may not be required.

Trial design:

"The current guideline does not tackle the issue of an alternative to the equivalence design, which is currently expected for clinical studies to show sufficient similarity in efficacy (mentioned e.g. in the Draft Guideline on Similar Biological Medicinal Products Containing Monoclonal Antibodies).

In the equivalence trial clinically meaningful differences including superior efficacy between the biosimilar and the reference product are excluded if the 95 % confidence interval of the treatment difference is fully contained within the pre-specified two-sided (upper and lower) comparability margins.

In the case of a non-inferiority study, only the lower margin is defined, i.e. the statistical analysis can only show that the product in not inferior to the reference product, but cannot exclude the possibility that the biosimilar product would be better than the

[52] Neutrophil:
It refers to a neutrophilic white blood cell, according to the Oxford Online Dictionary.

reference product. The number of patients needed for an equivalence trial tends to be larger than that for the non-inferiority trial.
Also the possibility of using a non-inferiority design in the pivotal phase III studies in certain cases will be considered".

Conclusion/Comment:

- The trial design also impacts the size. In principle, there are two different designs: non inferiority design and equivalence trials design. Ema does not exclude any of them from being carried out.

Phase III:

"The BMWP will discuss, in case relevant markers are not available and similarity has been shown with selected PK parameters, whether the development of a biosimilar product can proceed to the pivotal phase III studies.
With regard to the design of the phase III studies, the BMWP intends to update the guideline to include examples of surrogate endpoints which are being already recommended in various other biosimilar guidelines, e.g. in the Draft Guideline on Similar Biological Medicinal Products Containing Monoclonal Antibodies. The BMWP will discuss on how to further define considerations around acceptance of pharmacodynamics markers for demonstration of clinical similarity, and how to generate sufficient safety data in addition".

Conclusion/Comment:

- Phase III is mentioned but phase II is not.

Phase I:

"With regard to the phase I studies the guideline recommends the selection of relevant pharmacodynamic markers, but such markers are not always available and/or cannot be shown to
reflect efficacy. Many biological products e.g. monoclonal antibodies, have several approved indications, and the principle emerges that the most sensitive one of them should be chosen to show therapeutic similarity".

Conclusion/Comment:

- This means that it is most sensible for the efficacy to be done in phase III trials.

Extrapolation:

For extrapolation of indication the following was mentioned:

"When efficacy and safety has been shown in one indication, extrapolation to the other indications is already now possible under certain circumstances.EMA/CHMP/BMWP/572828/2011 3/5. The BMWP intends to discuss these criteria and potentially revise or amend them with the experience gained so far".

"The extrapolation to other indications which are not investigated during the clinical development program of the biosimilar product will be discussed".

Conclusion/Comment:

- EMA is not fully sure about extrapolation criteria and awaits more (if any happened so far) cases of extrapolation. Thus showing the attitude of the EMA with regard to extrapolation is adaptive.

Table: 10
Summary of Concept paper on the revision of the guideline on similar biological medicinal products containing biotechnology derived proteins as active substance: Non-clinical and clinical issues EMA/CHMP/BMWP/572828/2011 (56) **with regard to the research question** of this study.

Research question	**Input from the guide**
What are the implications of heterogeneity and variation through the life cycle of the biosimilar and the reference biologic?	Quality of the biosimilar
Does the Biosimilar need to achieve a quality profile which falls within the quality profile of the reference biologic? Under which circumstances could this approach be deviated from?	
What is proposed if the reference biologic changes its quality profile during the biosimilar development program?	
What should be the scope of trials?	Trial design Phase III Phase I Trial size Scope: A new concept is available for the scope of clinical trials
Extrapolation	Extrapolation

2.4. Reference to other biosimilar regulations (for informational purposes only)

Here are some insights into the state of biosimilar regulation in the US and Japan. Further, the WHO has released a guidance document for biosimilars in 2009. These guidance documents were not further followed up within this study, as they are outside the scope of the research question.

2.4.1 The US

As of Feb 9, 2012, the **FDA issued the draft guidance on biosimilar product development.** The biosimilar approval pathway was set by the Biologics Price Competition and Innovation Act of 2009 (BPCIA). Patient Protection and Affordable Care Act, 2010, amended the Public Health Service Act, to create an abbreviated approval pathway. This pathway is included under this Act section 351(k) for biological products that are demonstrated to be highly similar (biosimilar) or to be interchangeable with an FDA-licensed biological product. Currently the FDA made public two guidance documents, which are in a draft status.

> *"Scientific Considerations in Demonstrating Biosimilarity to a Reference Product:*
>
> *"The draft guidance is intended to assist companies in demonstrating that a proposed therapeutic protein product is biosimilar to a reference product for the purpose of submitting an application, called a "351(k)" application, to the FDA. This draft guidance describes a risk-based "totality-of-the-evidence" approach that the FDA intends to use to evaluate the data and information submitted in support of a determination of biosimilarity of the proposed product to the reference product. As outlined in the draft guidance, FDA recommends a stepwise approach in the development of biosimilar products.*
>
> *"Quality Considerations in Demonstrating Biosimilarity to a Reference Protein Product: The draft guidance provides an overview of analytical factors to consider when assessing biosimilarity between a proposed therapeutic protein product and a reference product for the purpose of submitting a 351(k) application. This includes the importance of extensive analytical, physico-chemical and biological characterization in demonstrating that the proposed biosimilar product is highly similar to the reference product notwithstanding minor differences in clinically inactive components.*
>
> *"Biosimilars: Questions and Answers Regarding Implementation of the Biologics Price Competition and Innovation Act of 2009:*
> *The draft guidance provides answers to common questions from people interested in developing biosimilar products. The question and answer format addresses questions that may arise in the early stages of product development, such as how to request*

meetings with the FDA, addressing differences in formulation from the reference product, how to request exclusivity, and other topics". (57)

The debate about the draft guides issued by the FDA is about the following topics:

1. Interchangeably of the reference biologic and the biosimilar. The burden is considered by the FDA, as well as industry, to be considerably high, to prove interchange ability.
2. There are user fees to pay by the biosimilar industry for the examination of an application.
3. If clinical trials will be required for all biosimilars, or some very simply biologics might not require them. (58)

2.4.2 Japan

Japan's Ministry of Health, Labour and Welfare (MHLW) issues a series of guidance documents:

1. "Guidelines for the Quality, Safety and Efficacy Assurance of follow-on biologics" (Yakushoku shinsahatu 0304007 by MHLW / March 4, 2009)
2. "Nonproprietary name and brand name of follow-on biologics" (Yakushoku shinsahatu 0304011 by MHLW / March 4, 2009)
3. "Revision of marketing approval application" (Yakushoku shinsahatu 0331015 by MHLW / March 4, 2009)

(59, 60)
Japan's biosimilar guidelines follow a similarity approach comparable to the one in the EU. The MHLW defines a biosimilar as: *"a biotechnological product that is produced by a subsequent-entry manufacturer and claimed to be comparable to a biopharmaceutical product already approved in Japan".*
Clinical studies are required to verify the efficacy and safety of the biosimilars and to prove bioequivalence. Extrapolation of indication is permissible if, *"similar pharmacological action can be expected in other indications through relevant pharmacokinetics and pharmacodynamics studies or analysis".* (61)

2.4.3 Canada

Health Canada issued a Guidance for Sponsors: Information and Submission Requirements for Subsequent Entry Biologics (SEBs) as a frame work, for all biologics seeking an abbreviated approval process based on an already authorized biologic, where similarity will be required to be shown/ demonstrated.

The following criteria are used to define SEB qualification:

- *"A suitable reference biologic drug exists that: a) was originally authorized for sale based on a complete data package; and b) has significant safety and efficacy data accumulated such that the demonstration of similarity will bring into relevance a substantial body of reliable data;*

- *the product (i.e., SEB) can be well characterized by a set of modern analytical methods; and*
- *the SEB, through extensive characterization and analysis, can be judged similar to the reference biologic drug by meeting an appropriate set of pre-determined criteria".*

The SEB concept relies in part, on prior information regarding the original biologic drug in order to present a reduced clinical and non-clinical package as part of the submission. (62)

2.4.4 The WHO

The WHO's guide for biosimilars recommends that head-to-head comparisons are carried out between the biosimilar and the reference biologic. The same reference biologic should be used throughout the entire comparability exercise.

1. *"The manufacturer needs to demonstrate that the chosen REFERENCE BIOLOGIC is suitable to support the application for marketing authorization of a BIOSIMILAR.*
2. *The REFERENCE BIOLOGIC should be licensed based on a full quality, safety, and efficacy data. Therefore a BIOSIMILAR should not be considered as a choice for REFERENCE BIOLOGIC.*
3. *The same REFERENCE BIOLOGIC should be used throughout the development of the BIOSIMILAR (i.e. for the comparative quality, non-clinical, and clinical studies).*
4. *The drug substance of the REFERENCE BIOLOGIC and the BIOSIMILAR must be shown to be similar.*
5. *The dosage form and route of administration of the BIOSIMILAR should be the same as that of the REFERENCE BIOLOGIC.*

"The REFERENCE BIOLOGIC should be licensed and widely marketed in another jurisdiction which has well-established regulatory framework and principles, as well as considerable experience of evaluation of bio therapeutic products, and post-marketing surveillance activities.

"The acceptance of a REFERENCE BIOLOGIC for evaluation of a BIOSIMILAR in a country does not imply approval for use of the REFERENCE BIOLOGIC by the NRA of that country."

Further, the WHO acknowledges the, *"heterogeneous nature of proteins (especially those with extensive post-translational modifications such as glycoproteins)"*, and the issues associated with the impact of quality profile variation and the sometimes unknown effect of that on the safety and efficacy.

"For example, if differences are found in glycosylation patterns that alter the biodistribution of the

product and thereby change the dosing scheme, then this product cannot be considered a BIOSIMILAR.

"Other differences between the BIOSIMILAR and REFERENCE BIOLOGIC may be acceptable, and would not trigger the need for extra non-clinical and/or clinical evaluation. For example, a therapeutic protein that has lower levels of protein aggregates would, in most cases, be predicted to have a better safety profile than the REFERENCE BIOLOGIC and would not need added clinical evaluation. Along the same lines, if heterogeneity in the terminal amino acids of the REFERENCE BIOLOGIC is known, and sufficiently documented, without affecting the bioactivity, distribution, or immunogenicity of the RERFENCE BIOLOGIC or similar products in its class, then there may be no need for added clinical safety or efficacy studies based upon this heterogeneity of the RPB and BIOSIMILAR." (1)

Chapter 3: Materials and Methods

The choice of materials and methods employed resulted from practical considerations, namely the knowledge and awareness of the tools to use, the suitability to tackle my research questions and funding.

Despite no specialized materials being utilized to conduct this work, the following might be worth mentioning:

1. As a guidance for the overall writing process of the study, after considering others like Randolph and Justus (2009) (63), the following source was used:

    i. *How to Write a Thesis: A Working Guide*
    ii. *R Chandrasekhar Australian Research Centre for Medical Engineering (ARCME)*
    iii. *The University of Western Australia* chandra@arcme.uwa.edu.au
    iv. *Written: 24 Feb 2000; Revised: 30 April 2002* (64).

2. For referencing and bibliography "Mendeley library" software was regularly used together with a Word program plug-in: http://www.mendeley.com
3. For the study standard office software from Microsoft was employed: *Microsoft Office Home and Student 2010.*
4. For communication and e-mailing, a standard Gmail account was employed: *Google Mail - Gmail – Google.*

5. For general information and research purposes Google was employed, however Google was never employed to answer to any particular research questions. Google is considered by the author as a good tool to obtain a quick over view whenever required.

6. As guidance for the design of the questionnaire the following documents were employed:
    i. *How to design a questionnaire*
    ii. *Wai-Ching Leung has some practical advice on questionnaires*
    iii. *STUDENTBMJ VOLUME 9 JUNE 2001 studentbmj.com* (65)
    iv. and
    v. *A guide to... ...Questionnaire Design* (66).

7. For questionnaire 1.1, an online survey, Survey Monkey software/website was used: *de.surveymonkey.com/.* This one was elected among others such as: surveygizmo.com/ and www.statpac.com/ due is easy to handle setup options.

3.1 Methods used and rational for choosing them

In total three different research methods were employed in this study. The selection and identification process of possible and suitable research methods was interconnected with the selection and identification process of suitable and meaningful research questions, refer to section: 1.0. Rationale on the selection of the topic (as guide for future students).

A mixture of methods (such as surveys, submissions to the EMA and literature research) were used for the research and have been beneficial in revealing new and meaningful information. In the current study, through the employment of active research means (via questionnaires and submissions of questions/comments to EMA), the result will contain relevant conclusions related to current and future issues in the field of biosimilars, as discussed in section: 1.2 Statement of the main problem, subsequent research questions and test-functions.

3.1.1 Questionnaires and submissions

For the purpose of this study, two questionnaires were employed:

1) Questionnaire 1.1 was an online questionnaire (or online-survey) that was posted with a link in dedicated biosimilar forums. In order to visualise the questionnaire itself refer to: Appendix: 3 Questionnaire 1.1 and refer to sections: 3.2 Rationale for using the employed research methodologies and 3.3 Practical aspects for more detail on the questionnaire.

2) Questionnaire 1.2 was a Word document that was sent via e-mail to specific experts in the field of biosimilars. For more details refer to: Appendix: 4 Questionnaire 1.2 for the questionnaire and refer to sections: 3.2 Rationale for using the employed research methodologies and 3.3 Practical aspects for more detail on the questionnaire.

For the purpose of this study two sets of questions and comments were submitted to EMA:

1) Refer to: Appendix: 1 Questions and comments submitted to the EMA, for questions and comments on the *"Guideline on similar biological medicinal products containing monoclonal antibodies Draft"*.
2) Refer to: Appendix: 2 Questions and comments submitted to the EMA, for questions and comments on the *"Guideline on similar biological medicinal products containing biotechnology-derived proteins as active substance: quality issues (revision 1) Draft"*.

3.1.2 Literature review

For this study, a multilayer literature review was conducted that allowed the capture of systematic information and could possibly reveal new insights, once the information was analyzed. Dedicated keywords were employed and were carefully selected to obtain specific information on the research questions. For keywords employed see the relevant text section, figure or table. Whenever possible, the source papers and articles were downloaded, refer also to: Bibliography or References. The main sources consulted were the following:

1) EMA website, including EMA guidance documents on biosimilars (30)
http://www.ema.europa.eu/ema/index.jsp?curl=pages/regulation/general/general_content_000408.jsp&mid=WC0b01ac058002958c

2) Pubmed (31)
http://www.ncbi.nlm.nih.gov/pubmed/

3) Gabionline (32)
http://www.gabionline.net/

3.2 Rationale for using the employed research methodologies

The reason for the choice of the research methodologies was primary linked to the research field and the research questions; thus, experiments with appropriate controls to test a theory or hypothesis were not applicable in the case of this study. Variables used regularly in a scientific setting cannot be tested independently from each other, nor is there the possibility to work on a laboratory model. Hence, the focus of this research was more on a qualitative rather than a quantitative matter.

Given the constraints of this study, it was not always possible to properly summarize and statistically evaluate, as would be in the natural science setting, data points. It would be (of course) preferable if there was high quantities of discovered data. However, it is unlikely that statistically relevant sample sizes, for example, in the case of responses to: Appendix: 3 Questionnaire 1.1 and Appendix: 4 Questionnaire 1.2, will be collected or even achievable.

This is the main limitation of the chosen field of research, and is related to the research questions themselves, that is to say, making the available "population"[53] (statistical term) of participants rather limited.

Therefore, the main part of this research was focused on qualitative research, i.e. in depth analysis of particular and technically challenging aspects of biosimilars. The research methodologies employed tackle those aspects by addressing a circle of stakeholders, which is directly involved with the legistration and guidance creating/drafting for biosimilars with their in-depth/expert knowledge.

[53] Population:
Statistics - a finite or infinite collection of items under consideration. Oxford Online Dictionary.

This approach is also continued and reflected in the detailed and partially technical questions in Appendix: 3 Questionnaire 1.1 and especially Appendix: 4 Questionnaire 1.2. The submissions to the EMA and their complex and detailed concepts reflect this detailed approach too.

However, it is imperative to acknowledge that the research methodologies were rather limited to the biosimilars community, including the EMA and industry. As a consequence, EMA as well as the Industry were directly approached. One reason for this direct approach (the use of submission and questionnaires) was that the research questions were directly linked to the guidance documents issued by the EMA. This guidance documents issuing process can be compared to the process when a new European directive/regulation is being discussed during its formation in Brussels. Politicians and "interested" parties consult, lobby and form alliances etc. to obtain their input in the legislation.

Concerning the EMA and the issuance of biosimilar guidance documents, there are parallel processes happening. Those are similar to the ones for "Brussels law making" with regard to lobbying and hearings, as those guidance documents define requirements that have financial impacts for the industry. For more details refer to the section: 1.4 Description of the European Regulatory Environment with Regard to Biosimilars

Thus, it was natural that during the elaboration of this study a consultation with those industry parties involved in this hearing process, next to the authority itself which is the EMA, was attempted. Despite being a rather direct research methodology, it turned out well and rather effective, as only the relevant stakeholders were integrated in this research.

When sending submissions to the EMA, with suggestions on particular draft guidance documents, as done in this study, the key advantage is that thoughts and concepts find direct access to the target group within the EMA: the BWP[54], which is in charge of draft guidance issuance and revision and with general responsibility for biosimilar matters, refer to section: 4.1 Introduction where comments for the *"Guideline on similar biological medicinal products containing monoclonal antibodies Draft"*, submitted on 30 May 2011, are included. The *"Guideline on similar biological medicinal products containing biotechnology-derived proteins as active substance: quality issues (revision 1) Draft"*, was released for public consultation on 31 May 2012. This example indicates the impact that questions and comments may have on future guidance.

3.2.1 General suitability of the research methods employed

With regard to the research questions, the employed methods should be suitable to tackle and answer the research questions conclusively. Practical aspects occurring during the application

---

[54] The Biologics Working Party (BWP) provides recommendations to the European Medicines Agency's scientific committees on all matters relating directly or indirectly to quality and safety aspects relating to biological and biotechnological medicines.
http://www.ema.europa.eu/ema/index.jsp?curl=pages/contacts/CHMP/people_listing_000009.jsp&mid=WC0b01ac0580028d2c

phase of the chosen methods could undermine the quality of the research. Those aspects will be discussed in section: 3.3 Practical aspects.

Another limitation of this research was related to funding; there was no dedicated budget that resulted in a situation where the research methodologies faced limitations. This effect was reduced by the overall quality of the study, however, meant more time was spent shaping the qualitative research.

3.3 Practical aspects

3.3.1 Practical aspects of the questionnaires and critical considerations

Questionnaires are commonly used for many different research areas, such as to probe for public opinions or to question members of institutions about views on a certain topic. Assuming that statistical pre-conditions are met, the hypothesis can be found wrong or can be found "failed to reject", i.e. evidence suggests that the hypothesis is true rather than false.

A potential pitfall of using questionnaires in this study is that the number of suitable participants for the questionnaire is rather limited, thus the response rates were limited too.

The second major pitfall was that the participants may also have an agenda, which could be driven by financial interest of their employers, and is presumably true for participants from industry and for various types of consultants, who offer their services to industry. In order to avoid pitfalls and limit foreseeable problems, the following measures were taken:

- **The risk of obtaining answers form personages with low level knowledge was limited by:**

1. Posting questionnaire 1.1 in a dedicated forum for biosimilars and regulatory affairs where the participation of non-knowledgeable members is strictly limited. Refer to section: 3.3 Practical aspects for details.
2. Targeted contacting of possible participants over the phone whenever possible and subsequently e-mailing questionnaire 1.2 to the identified stakeholders based on an e-mail distribution lists of people who participated in the mAbs guide (67). The list included members of the pharmaceutical industry, specialised consultants, and members of competent authorities.
3. Targeted e-mailing to the EMA. The EMA, and especially BWP members, are specialized on biosimilars and were previously asked if they could complete the questionnaire.
4. Targeted e-mailing of the questionnaire to a series of competent authorities in the EU, whose members are also part of the EMA, especially the BWP, with responsibility for biosimilars.
5. Targeted contacting of world experts about biosimilar regulation, who participated in the drafting of the WHO guide for biosimilars. They were asked to fill in either Appendix: 3 Questionnaire 1.1 or Appendix: 4 Questionnaire 1.2.

As can be seen from the Figures: 4-9, this strategy was effective as the respondents were highly educated with the majority having a Masters degree or higher and they work in relevant areas, from their professional backgrounds.

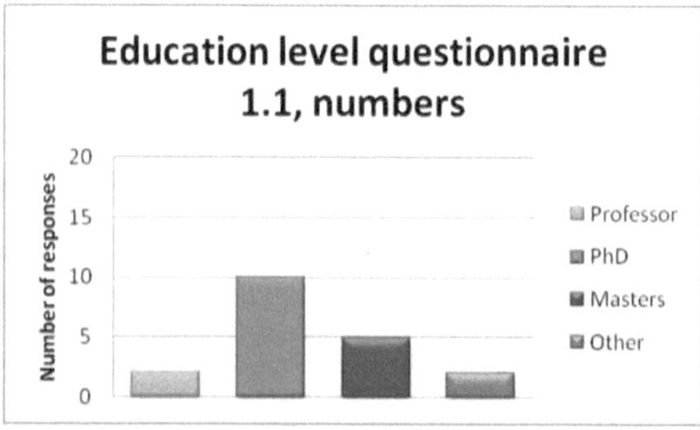

Figure: 4
Education level of participants taking questionnaire 1.1 (n=19).

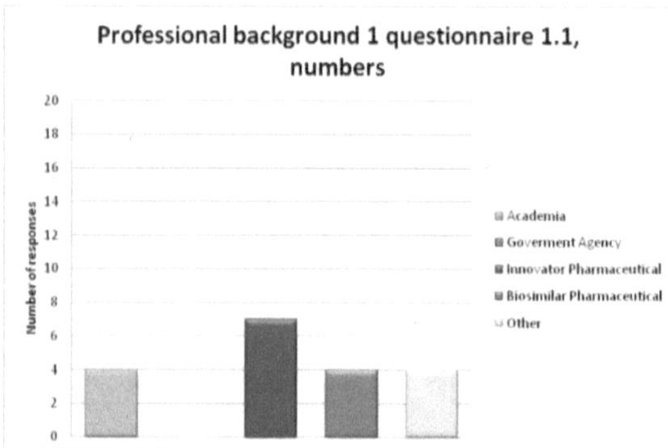

Figure: 5
Professional background (place of work) of participants taking questionnaire 1.1 (n=19).

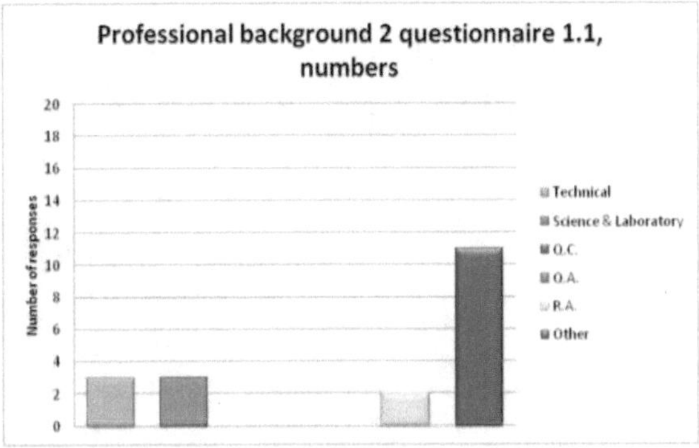

Figure: 6
Professional background (role) of participants taking questionnaire 1.1 (n=19).

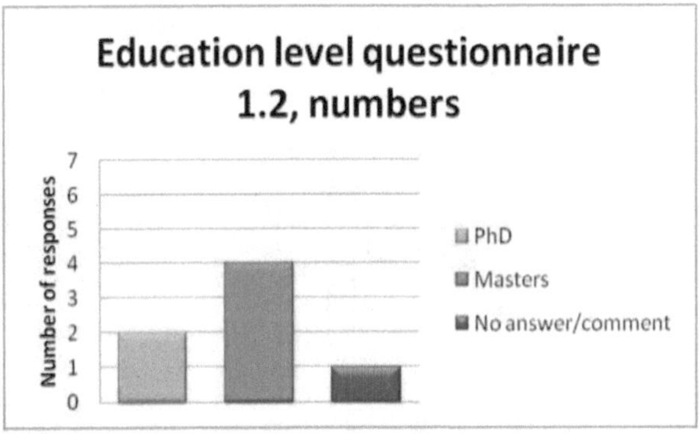

Figure: 7
Education level of participants taking questionnaire 1.2 (n=7).

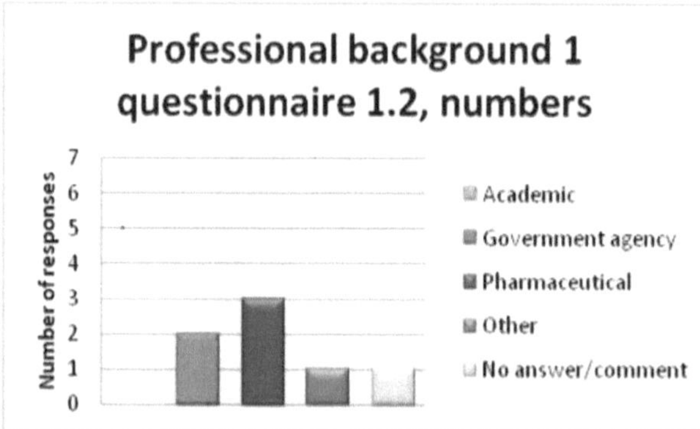

Figure: 8
Professional background (place of work) of participants taking questionnaire 1.2 (n=7).

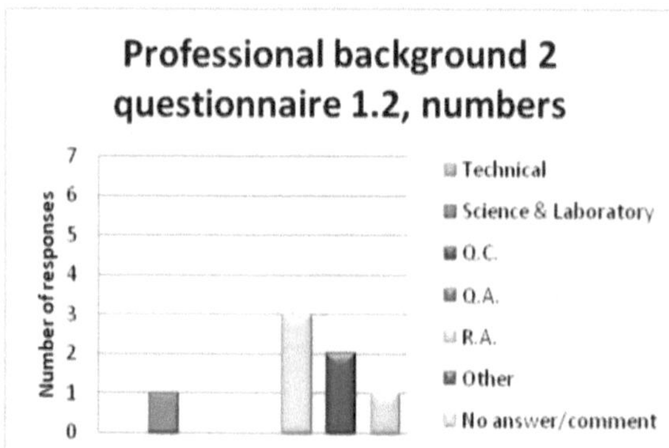

Figure: 9
Professional background (role) of participants taking questionnaire 1.2 (n=7).

- **Deliberately biased input was avoided by the design of the questionnaire:**

1. Both questionnaires contain a section assessing the knowledge and skill level of the respondent in conjunction with their professional background.

2. It is expected that various stakeholders have their respective agendas, in the case of industry, potentially influenced by financial considerations. Industry is equally heavily influenced through the consultation and lobbying process, which is associated with the biosimilar guidance regulation. Thus, their views are of equal importance to the healthcare authorities or experts in the field of biosimilars.

- **Good response rate vs. insufficient participants**

An insufficient number of participants is unavoidable, as the biosimilar community is "limited". It is therefore, expected that there will not be enough responses for statistical purposes.

1. Questionnaire 1.1 was publicly available online via a Survey Monkey system. This was intended for a higher participation rate. The number of questions is relatively small, with only 9 questions due to longer questionnaires being unpopular.

2. The survey monkey questionnaire 1.1 was first posted in the previously mentioned forum from the 15.02.2013 and was renewed on a weekly basis to assure good visibility and awareness within those forums:

Name of the LinkedIn forum	Size of the LinkedIn forum	Description of the LinkedIn forum	First posting in the LinkedIn forum	Re-posting in the LinkedIn forum	Total length of posting
Biosimilar/Follow-On	Members (462) on 21.02.2012,	LinkedIn forum that is dedicated to biosimilar questions	Initially posted on the 14.02.2013	Renewed on a weekly basis on the 21.02.2013 28.02.2012 07.03.2012	4 weeks
Biosimilars	Members (3,316) on 21.02.2012,	LinkedIn forum that is dedicated to biosimilar questions	Initially posted on the 14.02.2013	Renewed on a weekly basis on the 21.02.2013 28.02.2012 07.03.2012	4 weeks
Quality & Regulatory Network	Members (56,094) on 21.02.2012	LinkedIn forum that is dedicated to regulatory questions	Initially posted on the 14.02.2013	Renewed on a weekly basis on the 21.02.2013 28.02.2012 07.03.2012	4 weeks

The following introduction was posted on a weekly basis to obtain a sufficiently high response rate.

Title of the posting on LinkedIn:

"Fancy knowing your view on biosimilars. Quality profile? Trial size? Extrapolation?"

Text of the posting on LinkedIn:

"For my study at the D.I.T., I would like to invite you to share your view. *http://www.surveymonkey.com/s/H7QPJ7W* *The benefits of participating will be that you obtain some of the results. This is your OPPORTUNITY.* *Please contact me, if you have questions via LinkedIn.*

3. Some effort was made to contact potential participants individually via direct e-mail or even over the phone for questionnaire 1.2, see Appendix: 4.

4. To obtain a higher response rate, personal contact was deemed as having a more positive impact on the response rate of participants.

5. In order to boost response rate for participants, it was decided not to reveal their identity. This measure was taken in order to allow them to express their views freely and to avoid potential conflicts of interests. Due to this confidentiality aspect, a detailed list of contacted participants and organizations cannot be revealed nor can the identity of the participants.

6. As a reward for participation in questionnaire 1.1 or 1.2, the participants were offered the chance of receiving the results.

- **Too complicated or imprecise questions**

1. This risk was limited by offering the participants the ability to contact the author of this study in order to obtain a clarification of a particular question (contact details were provided in questionnaires 1.1 and 1.2). Furthermore, there was the opportunity to fill in a comment section for questionnaires 1.1 and 1.2.

2. The number of questions in questionnaire 1.1 was limited to nine, which should allow for focused answers.

- **Confidentiality and conflict of interest of participants**

1. As mentioned before, the participant's details will not be published as part of this study. They will be contained in a separate folder that only the author has access to and they will only be revealed upon request by the supervisors of the study.

- **Analysable results**

1. In order to obtain results that can be further analysed, the two questionnaires were mainly based on a series of YES/No questions that prompted a selective response. This was regarded as a simple and efficient question type that leads to clear responses and enables comprehensive segregation of participant's views.

2. There was the option to contact the author of this study to clarify the questions in the questionnaire or the questions and comments submitted to the EMA, if wanted by the respective party, as e-mail contact details were provided.

3. There was the option in both questionnaires to write comments beneath the answer and to comment on the question or complement the answer if the participants wish to do so.

3.3.2 Practical aspects of the submissions and critical considerations

In general, any opportunity to contact the EMA related to the research questions should be regarded as a very important one.

In the period of time during which this study was conducted, two submissions to high impact guidance documents could be made.

The *"Guideline on similar biological medicinal products containing monoclonal antibodies Draft"*, is of high impact because mAbs are estimated to be a key market for biosimilars. Further the *"Guideline on similar biological medicinal products containing monoclonal antibodies Draft"* is extensive in its details. Additionally, extrapolation of indication is explicitly mentioned, which is considerable given the size of (large) the molecules and their mode of operation, which is so far the most complex one.

The other guide, *"Guideline on similar biological medicinal products containing biotechnology-derived proteins as active substance: quality issues (revision 1) Draft"*, is fundamentally linked with general quality questions related to biosimilars.

The scope to ask questions was limited to the topic of the guide, and this did not correspond 100% to the research questions in the study. The EMA focused their answers on comments and questions that were directly linked to the scope of the respective draft guidance. However, they also indicated that some questions will be addressed in new upcoming guidance (*"Guideline on similar biological medicinal products containing biotechnology-derived proteins as active substance: quality issues (revision 1) Draft"*), when quality related question were asked by the author in relation to the *"Guideline on similar biological medicinal products containing monoclonal antibodies Draft"*, which focused on non clinical and clinical issues.

This type of research methodology was not optimal as the questions that could be asked had to be related to the guidance itself, in line with the contained ideas and wording.

The other constraint on the knowledge gained by the author of this study was that once knowledge had increased, the research questions shifted as well.

Thus for the period when the *"Guideline on similar biological medicinal products containing monoclonal antibodies Draft"* was available for public commenting and questioning (till 30 May 2011), the research questions of this study were not yet finalised at that stage nor was it clear if the EMA would give a response. This was only clear when on the 30 May 2012 the EMA document EMA/205886/2012 *"Overview of comments received on 'Guideline on similar biological medicinal products containing monoclonal antibodies"* became available, and the questions and comments from the author of this study were partially responded to. Before that, participants in the questions and comment sessions were mainly industry and biosimilars experts, not students like the author of this study.

- **EMA not responding to the research questions as they were out of their scope**

The submissions to the EMA were made with great attention to detail and providing a rational for the proposed changes or comments. The scope of the questions was related to the guide itself, and if not, then a link or connection was established with other guidance documents, which could be considered to be higher in the hierarchy to the draft of the guidance document open for questions and commenting.

The EMA has already responded to the submission of questions and comments for the *"Guideline on similar biological medicinal products containing monoclonal antibodies Draft"*, as part of the questions submitted to the EMA, refer to Q and A released on the 30 May *"Overview of comments received on 'Guideline on similar biological medicinal products containing monoclonal Antibodies'"*.

The EMA also kindly invited the author of this study to attend a workshop on the *"Guideline on similar biological medicinal products containing monoclonal antibodies Draft"* in the Canary Warf building.

- **EMA not responding in time for the study as they require time to review and to consult internally**

This was a risk for the second submission on the *"Guideline on similar biological medicinal products containing biotechnology-derived proteins as active substance: quality issues (revision 1) Draft"*. The submissions deadline was the 30th November 2012. Although there was no publically available time plan, based on the expertise gained with the first submission, it is possible the EMA will not publish the answers prior to the time when this study is required to be submitted.

3.3.3 Practical aspects of the literature research and critical considerations

- **Outdated papers**

Biosimilars as defined within the EU have been on the market since 2006. Thus, most relevant publications were issued since that date. Concerning the literature research papers, the first criteria used for the choice was with regard to the relevance to the research question, keeping in mind, whenever possible, that the most recent publications have priority.

- **Biased publications that were not peer reviewed or articles where conflict of interest exists due to the financial links of the author with the industry**

This is a constraint that is unavoidable. EMA members also published articles in journals; however, those did not fully answer the research questions.

Publications sponsored by industry might be biased; however, they were also considered as a key source of information.

- **Inefficient and falsified data through wrong and inefficient research parameters such as data source, keywords**

The research in the cited literature indexes was always carried out with the cited keywords in the mentioned libraries. The keywords are indicated in the study as a note or comment in figures or text sections.

Chapter 4.0: What are the implications of heterogeneity and variation through the life cycle of the biosimilar and the reference biologic, from a European perspective?

1.0 What are the implications of heterogeneity and variation through the life cycle of the biosimilar and the reference biologic, from a European perspective?

Directly associated to that main research question, are questions 1.1) and 1.2) that discus the impact of changes in the quality profile.

> **1.1 Taking into account the new amendment of EMA/CHMP/BWP/617111/2010, with emphasis on the life-cycle of the Biosimilar, does the Biosimilar need to achieve a quality profile which falls within the quality profile of the reference biologic? Under which circumstances could this approach be deviated from?**
>
> **1.2 What is proposed if the reference biologic changes its quality profile during the biosimilar development program?**

4.1 Introduction

The lifecycle of any medicinal drug, starts with the drug development, extents over process development, continues to product launch and ends with product discontinuation, refer also to section: 2.2 Life cycle in relation to heterogeneity and variation.

Potentially, the life-cycle spans long periods of time ca. 20 years, or in some cases much more. During the life-cycle, changes can be made to the manufacturing process, changes in raw materials etc. can and will likely occur. This can have an impact on the medicinal product.

One could describe those changes as dynamic changes, as at some point in time they occur more frequent and with stronger severity than at others.

During the lifecycle of a biologic (be it a reference biologic or be it a biosimilar), changes to the quality profile occur, when taking the long lifespan of approved medicinal products into account. However, as biologics are heavily influenced by their manufacturing process (which is unique for each biologic and not 100% copy-able) and due to their complex structure, changes to the quality profile are more likely than for traditional (chemical) medicinal products. They tend to vary and posses a natural heterogeneity and are therefore particularly sensitive to variation, refer to sections: 2.2.1 Heterogeneity of biologics (proteins only) and 2.2.2 Variation in the biotechnology processes, in this study, describing the sources of heterogeneity and variation as well as the lifecycle of biologics.

This poses a problem for the regulator (EMA) and also industry as they are required to regulate a dynamic situation. Dynamic, within the biosimilars context, is not only to be understood as dynamic by "special" and therefore controlled changes; dynamic within the biosimilars context, also means the normal or unnoticed variation that occurs during normal "routine" operations.

The result of this is that the quality profile (Q-profile) of the biologic varies permanently. One can consider that on a lot to lot basis, but also from a product basis, when the changes are significant enough, for example, when the manufacturing process was changed.

In this section the implications of that heterogeneity and variation will be discussed from the biosimilar perspective, taking into account that the reference biologic is undergoing dynamic changes. This results in an ever changing Q-profile of the reference biologic and the biosimilar drug, in their respective life cycles, with various magnitudes/impacts of the changes.

When considering the research question: "***What are the implications of heterogeneity and variation through the life cycle of the biosimilar and the reference biologic, from a European perspective?"*** the biosimilar industry as well as the regulator faces a dilemma caused by this dynamic.

1) The biosimilar industry has the challenge to define their target QTPP[55].
If they do not take enough samples of the reference biologic, the QTTP could be too limited. This in itself is challenging as once the decision is reached by the biosimilar developer to initiate the biosimilar development program, they can only rely on current and possibly future batches of the reference biologic for studies, but cannot take past variation into account. This in itself poses a problem as a certain number of samples and data are required to define the QTPP.

The EMA was clear in their guide that the biosimilar developer needs to understand the reference biologic including its permitted ranges (variation and heterogeneity) very well to assure that the result of the development process is a biosimilar and not a bio better. This is related to the research questions 1.1: "***Taking into account the new amendment of EMA/CHMP/BWP/617111/2010, with emphasis on the life-cycle of the Biosimilar, does the Biosimilar need to achieve a quality profile which falls within the quality profile of the reference biologic? Under which circumstances could this approach be deviated from?***"

If the biosimilar has less heterogeneity and variation it risks becoming a bio better, which is strictly considered as a new product and the biosimilar guidance and approval facilitation does not apply any more.

[55] QTPP: Quality Target Product Profile, refer also to ICH Q8 R2 for more detail

The biosimilar developer therefore requires guidance on the expectations from the EMA and the guide *"Guideline on similar biological medicinal products containing biotechnology-derived proteins as active substance: quality issues (revision 1) Draft"* with regard to the Q-profile expectations and to what extent the biosimilar is required to be similar to the reference biologic, when the dynamic nature of the reference biologic is taken into account.

(Note: This is particularly required for biosimilar developers, as unlike process changes, much data and knowledge to assess the impact of the change in line with process understanding, is not available to the biosimilar developer.)

The EMA is required to give meaningful input on the biosimilar development strategy when taking the lifecycle of the reference biologic and the biosimilar into account. The problem is that these guidance documents could still be improved by making it clearer to the reader that the quality profile of the biosimilar and the reference biologic need to both be characterised to the same extent and it shall be shown that the clinical impact of variation and heterogeneity is similar for both. The later clinical challenge requires more precise guidance from EMA.

2) This dynamic poses many practical problems. It is clear that the burden to adapt the process and make the biosimilar similar enough to its reference is on the biosimilar developer's side. However, it is not clear if and to what extent the dynamic of the reference biologic influences and possibly jeopardizes that process.

EMA should have policies in place to tackle the dynamic of the reference biologic, especially when the Q-profile for the product has shifted. This is related to the research question: ***"What is proposed if the reference biologic changes its quality profile during the biosimilar development program?"***

A biosimilar developer may face the situation where over the course of the biosimilar development program the reference biologic shifts in its Q-profile.

The impacts of those shifts may be classified twofold:

1. As improvement of the reference biologic, in line with continuous improvement efforts (ICH Q11[56]) over a properly change control/ i.e. pre-change approval by EMA, based on submitted data showing assessing the impact of the change and there is approval for that change by the EMA.
2. As an unnoticed shift of the Q-profile that did not trigger any safeguards or reporting systems (medicinal products are constantly monitored for reported adverse events and an unnoticed change could cause such a situation).

---

[56] The International Conference on Harmonisation of Technical Requirements for Registration of Pharmaceuticals for Human Use (ICH) is unique in bringing together the regulatory authorities and pharmaceutical industry of Europe, Japan and the US to discuss scientific and technical aspects of drug registration. - Development and Manufacture of Drug Substances Q11.- http://www.ich.org/

However, one could also argue that no lot is identical to another lot of a biologic and the release specification, which are of course justified with solid data by the innovator industry, may not be sufficiently linked to cQA[57], given that at the time when the M.A. was issued the notion of cQA and their link to specification did not exist and rather simple process validation strategies were employed. Thus normal lot to lot variation of the reference biologic, for example, for highly glycosylated proteins, could result in problems for the biosimilar developer during the comparability exercise.

The EMA as well as both industry branches are required to think about such a scenario. All parties have so far been reluctant to take a stand on that potential issue.

Comment on difficulties with the comparability exercise due to the dynamic of quality profiles:

Scientists avoid, for good reason, experiments without adequate references and controls, as the outcome will be difficult to interpret. Thus in any scientific setting it is usually the aim to evaluate data with a reference, meaning a control which is traceable (i.e. stable).

Is the difference between sample 1 and sample 2 due to shifts in sample 1 or due to shifts in sample 2 or due to the method itself?

Sometimes it is possible to include additional stable and traceable controls to help evaluate results, however, sometimes this is not possible.

For example, for some analytical techniques used to build up the Q-profile to introduce independent controls such mass spectrometry or chromatograms, adequate calibration against a traceable reference as a standard is required. (68) However in a comparative analysis, as is the case for biosimilars (i.e. the reference is a biologic itself), the reference is unfortunately not stable, it is dynamic.

If unsure about the cause of a detected difference, one should take the strategy to compare more than one lot of the biologic to more than one lot of the reference. This strategy gives greater certainty, as the summary of data can be statistically evaluated by tests such as T-tests [58]etc. These can reveal whenever the data generated by biosimilar lots and reference lots

---

[57] cQA: critical Quality Attribute, refer also to ICH Q8 R2 for more detail

[58] T-test:

A t-test is a statistical hypothesis test based on a test statistic whose sampling distribution is a t-distribution. Various t-tests, strictly speaking, are aimed at testing hypotheses about populations with normal probability distribution. However, statistical research has shown that t-tests often provide quite adequate results for non-normally distributed populations too.

The term "t-test" is often used in a narrower sense - it refers to a popular test aimed at testing the hypothesis that the population mean is equal to some value m (see also t-statistic).

The most popular t-tests are aimed at testing the following hypotheses:

belong to the same population. If they belong to the same population, then one would assume that the biosimilar Q-profile of a product basis falls within the quality profile of the reference biologic.

Still, difficulties remain with the overall suitability of a biologic as a reference for the whole comparability exercise. This poses the question whether an independent reference needs to be developed and provided, as is already the case for various international standards, by health authorities such as the WHO Expert Committee on Biological Standardization.

The new proposed guide *"Guideline on similar biological medicinal products containing biotechnology-derived proteins as active substance: quality issues (revision 1)"* acknowledges the development and commercial life cycle of the biosimilar. The possible caveats with heterogeneity and variation throughout the lifecycle are explained in section: 2.2 Life cycle in relation to heterogeneity and variation.

1) Therefore the new proposed guide *"Guideline on similar biological medicinal products containing biotechnology-derived proteins as active substance: quality issues (revision 1)"* does not outline clearly enough that the aim for the biosimilar developer is to focus its development efforts on mimicking the Q-profile of the reference biologic, as consistency is mentioned in the draft guide, refer to Appendix: 1, it is required to clarify that consistency (which can be viewed as homogeneity and therefore as an extension of variation, is to be viewed as the same as consistency as the reference biologic). This does not mean that a development aim should be consistency itself, as this bears the risk that due to improvements in the purification area, that the biosimilar will be more consistent than the reference biologic. That cannot be the aim of the biosimilar development program. The EMA was asked to clarify this point, refer to Appendix: 1 and Appendix: 5.

2) The other point is also related to the development strategy. The Q-profile of the biosimilar can be different from the Q-profile of the reference biologic under certain conditions. The Q-profile of the reference biologic shall lay within the Q-profile of the reference biologic. The Q-profile of the reference biologic can be more homogenous, which might imply less impurities under the conditions that the clinical effects fulfil the biosimilar criteria. This is usually unknown at the beginning of the biosimilar development. When more clinical data becomes available this can be tackled better.

Postulate: The quality profile of the Biosimilar shall fall within the quality profile of the reference medicinal product. Deviation from that is permitted as long as safety and efficacy are not impacted.

---

1) The population mean is as hypothesized (the population variance is not known).

2) The means of two populations are equal (the population variances are not known but equal).

3) The means of two populations are equal (the population variances are not known and not equal).

4) The correlation coefficient for two random variables is zero.

5) The slope of the population regression line is zero.

Source: http://www.statistics.com/index.php?page=glossary&term_id=672

4.1.1 Why is the dynamic of the Q-profile for biologics of relevance?

There is potentially a great correlation with similarity in the quality profile and the success accomplishment of the clinical comparability i.e. the outcome of clinical safety and efficacy trials, due to immunogenicity or the ratio of isoform, for example.

The question is how often does the quality profile of the reference biologic drift and what are the implications of that for the biosimilar during the development phase of the biosimilar, where it is clearly required to achieve similarity? Whereas drift of the quality profile, is regarded as a change of the quality profile beyond the previously established limits; thus, it falls outside the established ranges of the reference biologic, as it can be set by analysis of as many lots as possible of the reference biologic during a given timeframe (by the biosimilar company).

EMA acknowledges this problem partially as it mentions in its guide for quality that once the M.A. is issued for the biosimilar, the Q-profiles of the reference biologic and the biosimilar are allowed to drift from each other (which is probably meant to be on a product basis); however, there is no indication given from the EMA about what happens if the reference biologic shifts during the development cycle of the biosimilar.

So far my understanding based on the new proposed guide *"Guideline on similar biological medicinal products containing biotechnology-derived proteins as active substance: quality issues (revision 1)"* is that the biosimilar developer is required to achieve the same quality profile as the reference biologic, this could mean that the development of the biosimilar in such a situation could face a significant drawback, resulting in loss of many years of development work and the associated financial resources.

Any further clarification of the described constrains is of relevance due to their impact on industry and the EMA. Thus this research question 1.0 and associated questions 1.1 and 1.2 are of utmost relevance.

4.2 Experimental procedure (methods and materials) employed

4.2.1. Literature research

1) The websites of the European Medicines Agency (EMA) were systematically consulted and selected biosimilar guidance documents were analyzed, refer to section: 3.3.1 Practical aspects of the questionnaires and critical considerations for details.

2) Publications were searched for in MEDLINE (Pubmed) and Gabi-online. Key search words were combinations of the words: biosimilar and heterogeneity or variation or life-cycle. Searches were limited to full-text English-language articles.

4.2.2. Questions and comments submitted to the EMA

Questions and comments were submitted to EMA utilising the provided form and contact details, refer to Appendix: 1 Questions and comments submitted to the EMA and to Appendix: 2 Questions and comments submitted to the EMA. The tables below represent extracts of those questions and comments, which are directly applicable to the research questions 1.0, 1.1 and 1.2 investigated in this chapter.

Note that the suitability of the comments or questions with regard to addressing the research questions can be better assessed when reading the whole context available in the appendices.

Table: 11
Extracts from questions and comments submitted to the EMA. Refer to Appendix: 1 Questions and comments submitted to the EMA for more details.
The questions and comments for the *"Guideline on similar biological medicinal products containing monoclonal antibodies"* were submitted on the 30.05.11 and the EMA provided responses to those questions and comments, refer to *"Overview of comments received on 'Guideline on similar biological medicinal products containing monoclonal antibodies"* (69).

Question and comments submitted to EMA extracted from Appendix: 1	**Notes and clarification in relation to research question**
1.) Therefore the variability of the Reference Medicinal Product itself (as a variable) leads to questions of how to fulfil requirements for the Biosimilar i.e. to demonstrate comparable clinical efficacy and safety versus the Reference Medicinal Product. Or differently expressed: 2.) When requiring any direct comparison, be it in a non-clinical or clinical setting, batch to batch variation of the Reference Medicinal Product (due to the glycosylation heterogeneity of mAbs or process impurities associated with the production of mAbs i.e. the full quality profile of the Reference Medicinal Product) and the quality profile of the Biosimilar need to be taken into account when designing studies and analyzing data. I call this train of thoughts in this submission the "overlaying issue of product heterogeneity and variation (1)".	Dynamic of the quality profiles and its implication (question 1.0)

Depending on assay sensitivity and methodology, comparative in vitro studies will always reveal differences in parameters cited in lines 135 to 138 of the draft guide. This is based on the reasonable assumption (based on available publications), that even the Reference Medicinal Product itself will vary between every batch, in for example the ratio glycosylated proteins and the level of glycosylation, refer also to [(1)]. Therefore the cited parameters will vary (slightly) for each batch of the Reference Medicinal Product or the Biosimilar, (again assuming sufficiently high assay sensitivity and methodology).	Should the quality profile of the biosimilar fall within the quality profile of the reference? (question 1.1)
A statement at the beginning of section 4.0 should incorporate the following **1.)** "the Biosimilar requires to such an extend similarity to the Reference Medicinal Product, as the Reference Medicinal Product is similar to itself, when compared on a batch to batch basis, as the optimum. Deviation of this approach is generally acceptable, but requires justification and possibly additional clinical evaluation in order to assess associated risks" Further to guide the industry and to avoid unnecessary costs and disappointments, it should be stated that: **2.)** "For non clinical setting it is advisable that the "Quality Profile", (refer to [(1)]) of the Biosimilar is within the levels of variation observed for the "Quality Profile" of the Reference Medicinal Product. Thoroughly determination of the "Quality Profile" of the Reference Medicinal Product can be used to set the margins of tolerance (for variation) for the Biosimilar "Quality Profile". Deviation of this approach is generally acceptable, but requires justification and possibly additional clinical evaluation in order to assess associated risks".	

Table: 12
Extracts from questions and comments submitted to the EMA for "Guideline on similar biological medicinal products containing biotechnology-derived proteins as active substance: quality issues (revision 1 Draft". **Refer to** Appendix: 2**, for more details. The questions and comments were submitted on the 28.11.2012 and the EMA has not published answers to these yet.**

Question and comments submitted to EMA, extract version	Notes and clarification in relation to research question
The Q-profile is based on characterization data (physicochemical and biological). Characterization methods have an error margins (as any assay/method) and one could add that margin to the context of variability/heterogeneity. However the link between to Q-profile of the biosimilar and the "comparability exercise" between biosimilar and reference biologic remains unclear! How does this relate to variability/heterogeneity and where is this incorporated into this guide?	Dynamic of the quality profiles and its implication (question 1.0)
Scenario 1: EMA permitted a process change for the reference biologic, which was considered acceptable. When compared head to head with the biosimilar in a clinical trial setting this change resulted in a measurable impact/difference (beyond the margins of equivalence). The Q-profile change of the reference biologic shifted, whereas the QTPP of the biosimilar was based on the pre-change Q-profile of the reference biologic. The post-change reference biologic had to be used for clinical trial comparability exercise due to normal shelf-life limitations. The biosimilar was shown to possess a better safety and efficacy profile compared to the reference biologic. • What is the consequence of that for the EMA? • What is the consequence of that for the reference product? • What is the consequence of that for the biosimilar? **Comment: This example was also verbally mentioned at the EMA in London by M.Osmane, in 2011 during a workshop with EMA about: "Guideline on similar biological medicinal products containing monoclonal antibodies – non-clinical and clinical issues", but as no official record is available for this question, this will not be further discussed in this study.** Scenario 2: EMA permitted a process change for the reference biologic, which was considered acceptable. When compared head to head with the biosimilar in a clinical trial setting this change resulted in a measurable impact/difference (beyond the margins of equivalence). The Q-profile change of the reference biologic shifted, whereas the QTPP of the biosimilar was based on the pre-change Q-profile of the reference biologic. The post-change reference biologic had to be used for clinical trial comparability exercise due to normal shelf-life limitations. The reference was	Implications in case where the reference biologic changes its quality profile, (question 1.2). Two scenarios where postulated.

shown to possess a better safety and efficacy profile • Is the comparability exercise considered failed? • Is refining of the QTPP required and some re-development?	
(I) Therefore a sentence at that stage (line 94) shall state that: "the initial or cardinal biosimilar development aim is to mimic the quality profile of the reference biologic drug substance including its variation/heterogeneity (on a lot basis*[1]) where possible."	The quality profile of the biosimilar should fall within the quality profile of the reference as outlined in I and II (question 1.1)
(II) Therefore EMA should state that for in section 4 line 107 for example: "the better the correlation of molecular characteristic (including process understanding) of the biosimilar drug substance with quality attributes is understood, the more the biosimilar developer can shift the focus (of this development strategy) from mimicking the Q-profile of the reference biologic including its variability/heterogeneity; towards manufacturing a more consistent/pure biosimilar drug substance and therefore eliminating some of the heterogeneity of the reference biologic, as long the safety and efficacy remain comparable within the margins of equivalence"	
(III) therefore EMA should state in section 4 that: "...at later development stages the established Q-attributes require confirmation in clinical trials and may be amended/completed, as the trial outcome cannot be foreseen. This could result in adjustment of the QTPP."	Dynamic of the quality profiles and its implication (question 1.0)

4.2.3 Survey research questions

A series of questions related to the research questions were made public, as a survey in selected forums as part of questionnaire 1.1 and was submitted as part of questionnaire 1.2 to selected stakeholders that included regulatory agencies, the innovator industry, the generics industry and consultants as well as other independent experts in the field.

The tables below represent extracts of those questions, which are directly applicable to the research questions 1.0, 1.1 and 1.2 investigated in this chapter; refer to Appendix: 3 and Appendix: 4 for details. Note that the suitability of the comments or questions with regard to addressing the research questions can be better assessed when reading the whole context available in the appendices.

Table: 13
Extracts from *"Questionnaire 1.1"*, refer to Appendix: 3, for more details.
The questions were published online in a dedicated forum, refer to section: 3.3.1 Practical aspects of the questionnaires and critical considerations.

Questions from survey monkey questionnaire 1.1	**Notes and relation to research question**
Do you think that multiple lots of the biosimilar against multiple lots of the reference biologic need to be part of any comparability exercise?	Research question 1.0
Should a biosimilar developer focus on mimicking the Q-profile of the reference biologic including all its variability OR should a biosimilar developer focus on homogeneity, for example, of the major isoform?	Research question 1.0 and 1.1
Do you think that multiple lots of the reference biologic need to be analysed, to understand the allowable limits for biosimilar manufacturing specification?	Research question 1.0 and 1.1
How much would you agree with the following sentence? "The quality profile of a biosimilar should fall within the limits of the quality profile of the innovator biologic, because one would expect that safety and efficacy will likely be similar too."	Research question 1.1
Would you agree with the following? "The initial or cardinal biosimilar development aim is to mimic the quality profile of the reference biologic drug substance including its variation/heterogeneity (on a lot basis) where possible." "The better the correlation of the molecular characteristic (including process understanding) of the biosimilar drug substance with quality attributes is understood, the more the biosimilar developer can shift the focus (of the development strategy) from mimicking the Q-profile of the reference biologic, including its variability/heterogeneity, towards manufacturing a more consistent/pure biosimilar drug substance; and therefore, eliminating some of the heterogeneity of the reference biologic, as long as the safety and efficacy remain comparable within the margins of equivalence." **Note:** **This comment was also submitted to the EMA, as part of the questions and comments for** ***"Guideline on similar biological medicinal products containing biotechnology-derived proteins as active substance: quality issues (revision 1)".***	The quality profile of the biosimilar should fall within the quality profile of the reference as outlined in table: 12 I and II (question 1.1)
Would you recommend to the biosimilar developer to initiate clinical trials (i.e. clinical comparability exercise), if the Q-profile of the reference biologic shifted and the QTPP of the biosimilar is based on the "old Q-profile" of the reference biologic. 1. Yes, as the risk of detecting a clinically significant difference due to	Implications in case where the reference biologic changes its quality profile,

changes of the Q-profile is a relatively low risk for the biosimilar developer. 2. Yes, as the "new Q-profile" of the reference biologic was approved by EMA and therefore should fall within similar safety and efficacy margins, as the "old Q-profile". Thus, it should not impact on the clinical comparability exercise. 3. Yes, as in the case where differences in clinical comparability exercises were detected in Q-profile of the biosimilar can be adjusted rapidly to follow the most recent Q-profile of the reference biologic. 4. No, as the Q-profile of the biosimilar should have prior clinical comparability exercises and a very similar Q-profile compared to the reference biologic. This should result in a similar clinical outcome. 5. No, as differences during the comparability beyond the margin or tolerance would require additional investigation and the burden of those is deemed as not desirable. 6. No, as the risk of detecting a clinically significant difference due to changes in the Q-profile is a relatively high risk for the biosimilar developer.	(question 1.2).

Table: 14
Extracts from *"Questionnaire 1.2"*, refer to Appendix: 4, for more details.
The questions were sent to dedicated experts, refer to section: 3.3.1 Practical aspects of the questionnaires and critical considerations.

Questions from questionnaire 1.2	**Notes and relation to research question**
Do you think that within one lot of a complex biologic (such as a mAb), it will always contain variations of mAb molecules (due to product or process related heterogeneity), which are slightly different to each other?	Research question 1.0
Do you think that the ratio of those slightly different (mAb)-molecules is unique for an individual lot?	Research question 1.0
Do you think that between multiple lots of a complex biologic (such as mAb), there will always be lot to lot variation of the Q-profile (due to product or process related variation and/or process changes)?	Research question 1.0
Do you think that multiple lots of the biosimilar against multiple lots of the reference biologic need to be part of any comparability exercise?	Research question 1.0
Do you think that multiple lots of the reference biologic need to be analyzed, to understand the allowable limits for biosimilar manufacturing specification?	Research question 1.0 and 1.1

What are the 5 major factors contributing to variation and heterogeneity of a biologic (such as mAb) in your opinion? Rated by significance.	Research question 1.0
Should a biosimilar developer focus on mimicking the Q-profile of the reference biologic including all its variability, Or focus on homogeneity, for example, of the major isoform? Please elaborate your thoughts and comment!	Research question 1.0 and 1.1
How much would you agree with the following sentence? "The quality profile of a biosimilar should fall within the limits of the quality profile of the innovator biologic, because one would expect that safety and efficacy will likely be similar too."	Research question 1.1
Would you recommend to the biosimilar developer to initiate clinical trials (i.e. clinical comparability exercise), if the Q-profile of the reference biologic shifted and the QTPP of the biosimilar is based on the "old Q-profile" of the reference biologic.	Research question 1.2

4.3 Results for main research questions 1.0 and directly associated research questions 1.1 and 1.2 that discus the impact of the dynamic to the quality profile

4.3.1 Results from the literature

4.3.1.1 Screening of regulatory literature (EMA) for research question series 1

The regulatory literature was screened as part of section: 2.3 Screening the above presented literature related to current biosimilar regulation with regard to the research questions.

The follow regulatory literature topics impact on the research questions series 1; presented as a summary in table form.

Table: 15 Extracts from section 2.3. series. Highlighting topics and findings from the analysis of regulatory documentation with regard to the research question. For the detailed text passage, please refer to section: 2.3. series and the following, which contains comments and further clarification.

Summary of regulatory literature review, impacting research question series 1 / Notes related to research question	**Refer to section**

Relates directly to research question 1.0, (the dynamic of the quality profiles), the initial guidance documents assumed that the same reference material throughout the comparability exercise can/should be used. This does not take the heterogeneity and variation due to changes (issue of dynamic of the quality profile) during the life cycle of both biologics into account. The second assumption is that "Highly purified products" will be compared during the comparability exercise.	2.3.1 Guideline on similar biological medicinal products CHMP/437/04 (49)
Related to research question 1.0, more recent guidance mentions again *"Highly purified proteins"* however it also acknowledges that *"biologics are complex mixtures"*. Thus taking the heterogeneity into account. However "batch to batch consistency" is recommended. The meaning of this is not clear (to M.Osmane). Note: If this means that batches/lots shall have a similar degree of heterogeneity and variation when compared to each other, this accommodates the dynamic of the quality profiles (research question 1.0) and gives guidance in line with research question 1.1. If consistency makes reference to lesser degree of heterogeneity and variation, then this can be interpreted in the context of "highly purified proteins" as a more pure product with less dynamic of the quality profile. This would result in it being a bio-better, but not a biosimilar anymore.	2.3.2 Guideline on comparability of medicinal products containing biotechnology derived proteins as active substance: Quality issues EMEA/CPMP/BWP/3207/00 (50)
Related to research question 1.0, it is outlined that refinements and clarification to the biosimilarity approach are needed.	2.3.4 Concept paper on the revision of the guideline on similar biological medicinal product EMA/CHMP/BMWP/572643/2011 (52)
Related to research question 1.0, the heterogeneity and variation is acknowledged as resulting in biosimilar purity and impurity profiles that will/may differ compared to the reference biologic. Again it is mentioned that the same reference material, should be used, however, a better description would be that similar reference material should be used, thus for the biosimilar the EMA acknowledges the heterogeneity and variation issues but does not seem to extend that notion clearly enough to the reference biologics. Related to research question 1.1, the line between	2.3.5 Guideline on similar biological medicinal products containing biotechnology-derived proteins as active substance: Quality issues EMEA/CHMP/BWP/49348/2005 (53)

product related substance and impurities is not fully clear, thus making it difficult to guide the biosimilar developer to what profile they shall develop the biosimilar with. Further, it is required to justify any differences in the quality profiles of the biosimilar compared to the reference biologic, in line with research question 1.1.	
Related to research question 1.0, the basis for the quality profile of the biosimilar is the combined scientific knowledge gained from the reference biologic, which results from analysis of multiple lots of the reference biologic. It is acknowledged and explicitly mentioned that the biosimilar has its own life-cycle, but no further information is given on the implications of that with the exception of mentioning that differences may be detected due to changes in the quality profile of the reference biologic, which may result in a situation, where the quality profile of the biosimilar requires adaptation. Related to research question 1.1, the usage of ICH Q8 R2 enforces together with the absence of mentioning purified or highly purified proteins, the view that the quality profile of the innovator sets the boundaries for the biosimilar developer. It is mentioned that the finished biosimilar should be highly similar to the reference; however, the degree of similarity, or better expressed as the range of permitted variability, requires justification.	2.3.6 Guideline on similar biological medicinal products containing biotechnology-derived proteins as active substance: Quality issues (revision 1) Draft EMA/CHMP/BWP/247713/2012 (54)
Related to research question 1.1, it is mentioned that there is a fine line between a biosimilar and a biobetter. This enforces the view that the quality profile of the reference biologic, including its impurities and heterogeneity and variation of the profiles (isoform profile) should be the basis for the biosimilar developer, as modern production techniques may result in reduced immunogenicity and the BMWP will discuss if these products would still qualify as a biosimilar.	2.3.8 Concept paper on the revision of the guideline on similar biological medicinal products containing biotechnology derived proteins as active substance: Non-clinical and clinical issues EMA/CHMP/BMWP/572828/2011 (56)

4.3.1.2 Screening of non regulatory literature and articles for research question series 1

An additional review of non-EMA papers and articles with regard to the research question was carried out, as outlined in section: 3.1.2 Literature review.

Related to the research questions 1.0 and 1.1:
Roger et al. brought to the attention of the public that improved analytical techniques may show that innovator products are not identical to the original biopharmaceutical that was licensed, years ago (19). In line with reports that analytical methods employed to characterise biologics are able to detect batch-to-batch differences in biologics, for example, the reference biologic such as mAbs, these same analytical methods should be applied to biosimilar mAbs (70).

This is confirmed by the answers to the questions employed in the questionnaires 1.1 and questionnaire 1.2 that the differences between the two batches of biologics are always present. If those were not detected, this reveals only that the employed analytical methods are not sensitive enough. It was emphasized that the question is not the ability to detect differences, but the determination of their clinical relevance (70).

The current European biosimilar regulatory pathway is restricted to biologics, but apparently not all copies of biologic molecules qualify. The EMEA CHMP guidelines state that comparability exercises to demonstrate similarity are more likely to be applied to highly purified products, which can be thoroughly characterized (e.g., biotech-derived medicinal products), than to other types of biologics. The implicit assumption is, therefore, that the pathway does not apply to more poorly purified biologics that are even more complex and difficult to characterize than a highly purified recombinant biologic.

This seems to contradict the rationale that biologic complexity and our inability to adequately characterize complex proteins, necessitates the comparability exercise (71).

The currently accepted approach to defining similarity is to integrate available information from orthogonal sources, e.g., physicochemical, binding, potency and (pre)clinical characterization. It is reported that similarity is difficult to describe, but after evaluating all the data, it is obvious (9). Biosimilar developers have the burden to show "*sufficient likeness to the originator product*" (19). The complexity of biopharmaceuticals makes it difficult to avoid heterogeneity between batches from the same manufacturing process and between the same proteins from different manufacturers (14).

Related to the research question 1.2:
In 2011 an important discussion was held with Dr. Kowid Ho (who kindly supported this study), which took the pervious articles into account and the stipulation that the quality profile over the entire life cycle of the reference biologic may well be dynamic.

> *"It was regarded as impossible to replicate exactly the innovator product, which is itself changing over time.*

Dr. Ho discussed issues and perspectives on the lifecycle of a biosimilar medicinal product and the reference product, which included aspects of product changes and resulting changes in quality attributes in line with ICH Q5E[59]*.*

Changes in quality attributes may be acceptable if it is demonstrated that there are no negative impact on safety and efficacy. Drift of quality attributes, the quality profile could occur over time,
Challenges for the biosimilar developer when defining the target variability of reference product attributes could occur.

Dr. Ho mentioned for acceptable changes, that depending on the quality attribute and their potential impact on safety and efficacy, could support and justify differences observed at the time of submission.

Dr. Ho also raised the question of whether biosimilars can really remain similar throughout their entire lifecycle as they will also have their own and independent lifecycle once they receive marketing authorization". (9)

There was one report suggesting that a defence strategy of the innovator biologic could be its complexity and the dynamic it brings, refer also to a comment made in section: 4.1 Introduction.

"In view of the major expense and short shelf-life of biologics, stockpiling the RMP for a long study is not practical. This example demonstrates just one of the difficulties faced by biosimilar trial
Sponsors trying to meet the requirements of regional RMP sourcing. Manufacturers of branded products are likely to use sophisticated defensive tactics, including the development of complex biopharmaceuticals, to maintain share." (25)

Possibly related to research question 1.2:
Declerck et al. pointed out that the first biosimilar Omnitrop was never compared in head to head trials with its reference as was suggested by the guidance documents. The reason for that was that during the development of the biosimilar the manufacturing process changed multiple times, thus the final version of Omnitrop was not tested in clinical trials in comparison with the reference biologic.

It is reported that alteration of the manufacturing process can have a significant impact on safety, as well as efficacy. For example, un-desired patient antibody levels detected with the biosimilar Somatropin decreased, after a site transfer within the same organisation (Sandoz)

[59] Q5E Comparability of Biotechnological/Biological Products Subject to Changes in their Manufacturing Process - http://www.ich.org/products/guidelines/quality/article/quality-guidelines.html

occurred. The formulation of interferon alpha 2a was shown to have a severe impact on safety i.e. immunogenicity (72).

In principle the same is applicable for mAb-biosimilars, where the glycosylation pattern influences heavily its bioactivity in its affinity towards other receptors such as Fc-gamma on different interacting cells, thus the efficacy of the mode of action of the mAb biosimilar is the same. It also influences the pharmacokinetics of mAbs, in its circulation time, for example (73). Therefore any change in a biologic product, whether an innovator product or a biosimilar, has a potential to impact immunogenicity (74), but also the overall safety and efficacy of a protein based biologic, especially in the case of large and glycosylated biologics with complex modes of actions such as mAbs.

It is remarkable that no head to head trials were carried out utilising the latest quality profiles of both biologics, taking the above outlined potential deficits into account.

4.3.2 Results from the questions and comments submitted to EMA

Refer to Appendix: 1, Appendix: 2 and Appendix: 5.

1) The questions and comments for the *"Guideline on similar biological medicinal products containing monoclonal antibodies"* were submitted on the 30.05.11 and the EMA provided responses to those questions and comments, refer to *"Overview of comments received on 'Guideline on similar biological medicinal products containing monoclonal antibodies"* (69) also represented in Appendix: 5.

2) Questions and comments for the *Guideline on similar biological medicinal products containing biotechnology-derived proteins as active substance: Quality issues (revision 1)* were submitted on the 28.11.2012. (A response from EMA is still outstanding in April 2013.)

3) Some comments made by M.Osmane for the *Guideline on similar biological medicinal products containing monoclonal antibodies,* refer to Appendix: 1 were incorporated into the *"Overview of comments received on 'Guideline on similar biological medicinal products containing monoclonal antibodies"* by the EMA.

Below is represented the incorporation of comments:
The Guideline on similar biological medicinal products containing biotechnology-derived proteins as active substance: Quality issues (revision 1) **released for public consultation 31.05.2012**, contains the following passage,

> *"The target acceptance criteria used in the comparability exercise should be justified. Quantitative limits should be established, where possible. <u>The relevance of these limits should be discussed, taking into account the number of reference medicinal product lots tested, the quality attribute investigated and the test method used. These limits should</u>*

not be wider than the range of variability of the representative reference medicinal product batches, unless otherwise justified"

Note: This passage or any similar text passage was not present in the current guide for GUIDELINE ON SIMILAR BIOLOGICAL MEDICINAL PRODUCTS CONTAINING BIOTECHNOLOGY-DERIVED PROTEINS AS ACTIVE SUBSTANCE: QUALITY ISSUES (EMEA/CHMP/BWP/49348/2005) and can be viewed as new.

The questions and comments **submitted on 30.05.11** the by M.Osmane for the *"Guideline on similar biological medicinal products containing monoclonal antibodies"*, refer to **Appendix: 4** were the following:

> *"The Biosimilar requires to such an extend similarity to the Reference Medicinal Product, as the Reference Medicinal Product is similar to itself, when compared on a batch to batch basis, as the optimum. Deviation of this approach is generally acceptable, but requires justification and possibly additional clinical evaluation in order to assess associated risks."*
>
> *"Further, to guide the industry and avoid unnecessary costs and disappointments, it should be stated that:*
>
> *For non clinical setting it is advisable that the "Quality Profile", (refer to [(1)]) of the Biosimilar is within the levels of variation observed for the "Quality Profile" of the Reference Medicinal Product.*
>
> *Thoroughly determination of the "Quality Profile" of the Reference Medicinal Product can be used to set the margins of tolerance (for variation) for the Biosimilar "Quality Profile". Deviation of this approach is generally acceptable, but requires justification and possibly additional clinical evaluation in order to assess associated risks".*

Note: As highlighted in the underlined text, there is a considerable adaptation from the side of the EMA with regard to this comment submitted in the currently proposed guidance document.

4.3.3 Results from the survey research method

1) The following findings were summarized from questionnaire 1.1. All comments from participants were cited.

Table: 16 Results from *"Questionnaire 1.1"*, refer to **Appendix: 3**, for more details. The questions were published online in a dedicated forum, refer to section: **3.3.1 Practical aspects of the questionnaires and critical considerations.**

Questions from survey monkey questionnaire 1.1	Results from research question
Do you think that multiple lots of the biosimilar against multiple lots of the reference biologic need to be part of any comparability exercise?	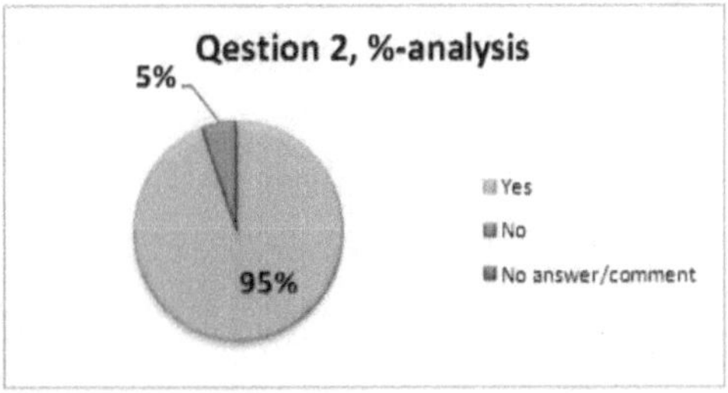 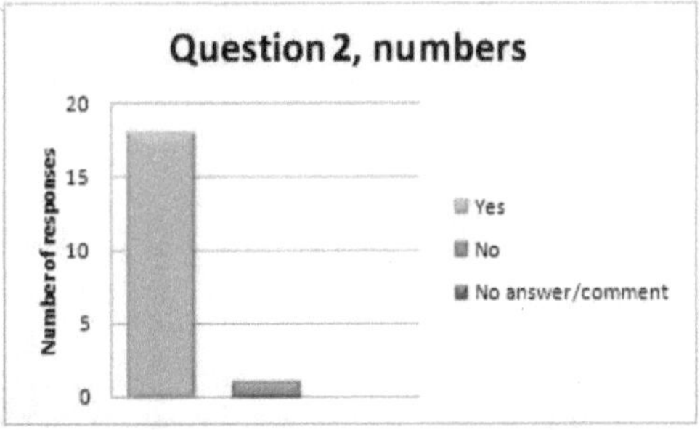 **Additional comments from participants:** • *Indeed, this is similar to the quality-by-design approach.* • *In a way. Multiple lots are required to understand the range of quality attributes from any process. Understanding this range for both the reference product and biosimilar product will be important in assessing how "similar" the two products are.* • *At least 5-10 to start program, and twice this number at registration. Attention should be paid to sampling strategy across globe for reference purchases.* • *Innovator will have lot-to-lot variability (process changes, minor other changes).*

<table>
<tr>
<td></td>
<td>

- *"Comparability" is totally the wrong term you want to be using. You apparently mean "biosimilarity." Comparability only refers to proving insignificant differences among different versions/iterations of an approved product.*

</td>
</tr>
<tr>
<td>Should a biosimilar developer focus on mimicking the Q-profile of the reference biologic including all its variability OR should a biosimilar developer focus on homogeneity, for example, of the major isoform?</td>
<td>

Research question 1.0 and 1.1

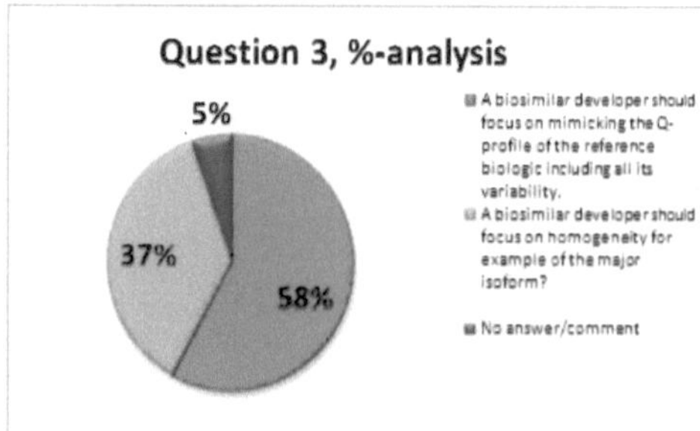

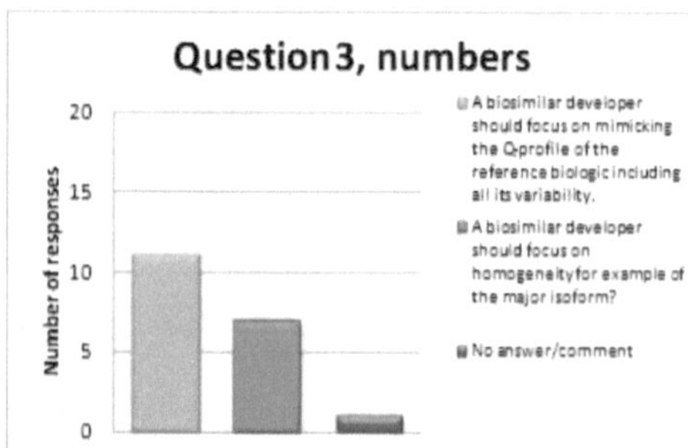

Additional comments from participants:

- *Better purity is a prime target for biosimilars, but leaving out biologically, particularly clinically, active "contaminants" risks not being proven biosimilar, e.g., if efficacy is too different.*
- *Either strategy can probably be rationalized and be okay as long as they generate the clinical data to show they have equivalent efficacy and no worse safety or immunogenicity*
- *Crazy anyway: they should copy even the contamination profile - only bio-better/bio-*

</td>
</tr>
</table>

	superior manuf. should divert from that concept, as they file a novel application / receive a new patent. • *BS company needs to pay attention to both aspects. BS company needs to show their process is both under control and at the same time is 'highly similar' to the reference.* • *All biologics are heterogeneous mixtures. If you were to only make the major isoform, it would not - by definition - be the same as a heterogeneous reference.* • *non-applicable* • *Differences in isoform composition may affect the properties (clinical as well as safety).* • *Either approach could be followed. as long as there is data substantiating the strategy followed* • *The biosimilar developer needs to develop a robust process that consistently produces product that falls within the Q-profile of the innovator product.*
Do you think that multiple lots of the reference biologic need to be analysed, to understand the allowable limits for biosimilar manufacturing specification?	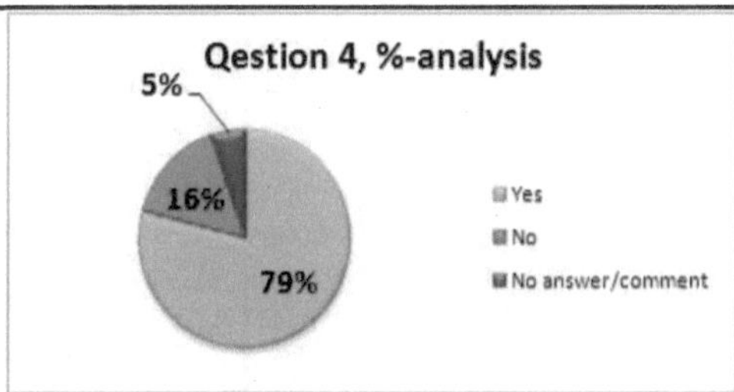 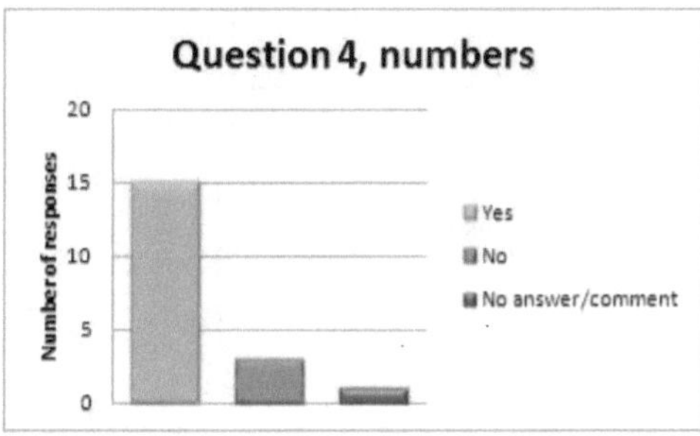

	Additional comments from participants: • *Indeed, this is similar to the quality-by-design approach.* • *In a way. Multiple lots are required to understand the range of quality attributes from any process. Understanding this range for both the reference product and biosimilar product will be important in assessing how "similar" the two products are.* • *At least 5-10 to start program, and twice this number at registration. Attention should be paid to sampling strategy across globe for reference purchases.* • *Innovator will have lot-to-lot variability (process changes, minor other changes).* • *"Comparability" is totally the wrong term you want to be using. You apparently mean "biosimilarity." Comparability only refers to proving insignificant differences among different versions/iterations of an approved product.* • *The limits of specification are part of a particular process as such and should not depend on variability observed in another product*
How much would you agree with the following sentence? "The quality profile of a biosimilar should fall within the limits of the quality profile of the innovator biologic, because one would expect that safety and efficacy will likely be similar too."	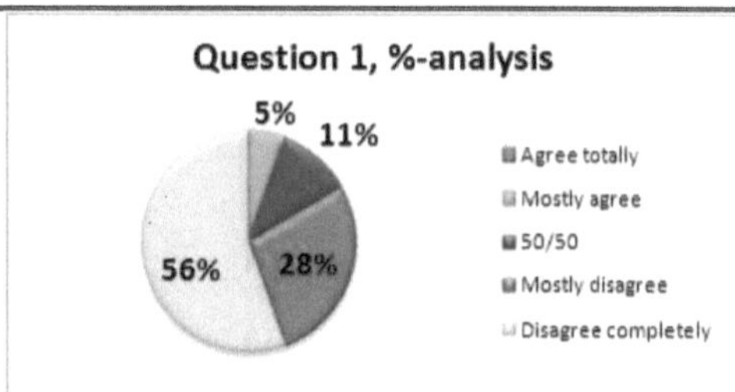

<table>
<tr>
<td></td>
<td>
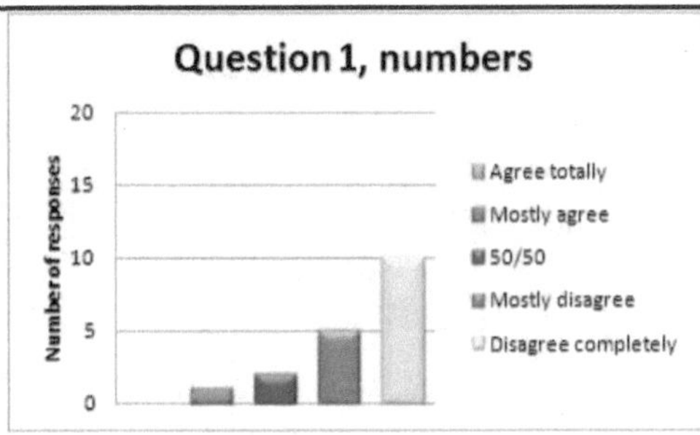

Additional comments from participants:

- *Ambiguous, quality profile implies the nature as well as the quantity of 'impurities'. The nature as well as the quantity of these may vary as long as it is acceptable from a 'pharmaceutical' point of view. In addition the manufacturer of the biosimilars has no access to the limits of specifications of the innovator product.*
- *Because the BS company will be using a different cell line and process, it would be expected that some QC measures would fall outside of the reference. BS company needs to explain the 'significance' if any of these excursions.*
- *They only need to be proven biosimilar or functionally, not analytically, equivalent (according to US law). In this context, the question is poorly constructed, with "should" being ambiguous - referring to requirements or preferences, it's unclear.*
- *There could be some minor differences which would still lead to a comparable efficacy and safety profile. It would be of advantage but I assume that depending on the quality attribute slight deviations might be acceptable -> case by case decision.*
</td>
</tr>
<tr>
<td>Would you agree with the following?

"The initial or cardinal biosimilar development aim is to mimic the</td>
<td>The quality profile of the biosimilar should fall within the quality profile of the reference as outlined in I and II (question 1.1).</td>
</tr>
</table>

quality profile of the reference biologic drug substance including its variation/heterogeneity (on a lot basis) where possible."

"The better the correlation of the molecular characteristic (including process understanding) of the biosimilar drug substance with quality attributes is understood, the more the biosimilar developer can shift the focus (of the development strategy) from mimicking the Q-profile of the reference biologic, including its variability/heterogeneity, towards manufacturing a more consistent/pure biosimilar drug substance; and therefore, eliminating some of the heterogeneity of the reference biologic, as long as the safety and efficacy remain comparable within the margins of equivalence."

Note:
This comment was also submitted to the EMA, as part of the questions and comments for *"Guideline on similar biological medicinal products containing biotechnology-derived proteins as active substance: quality issues (revision 1)"*

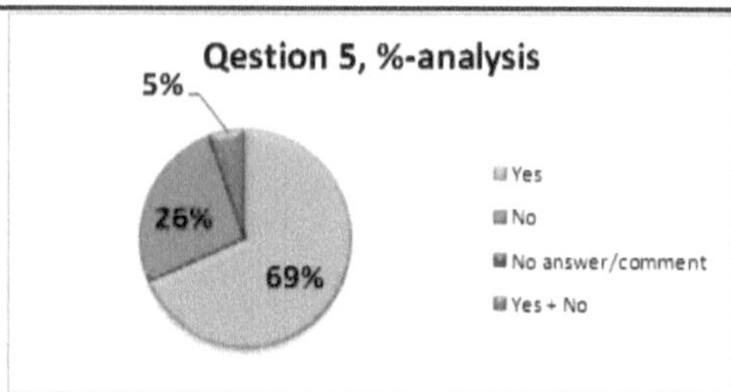

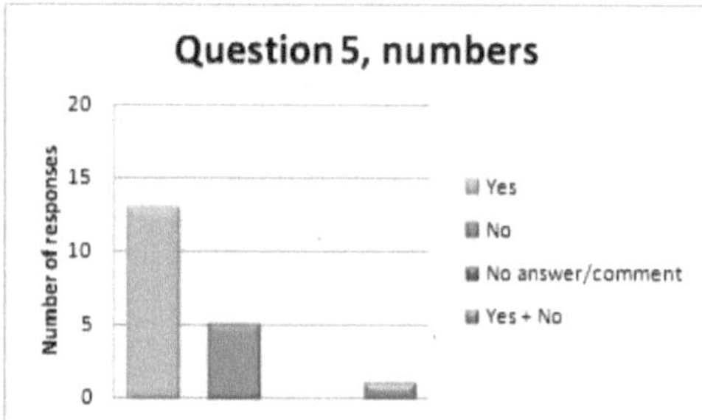

Additional comments from participants:

- *No, these appear two different (opposite !) statements. When deciding to go for a biosimilar, then one cannot try or focus to eliminate particular properties or characteristics that are part of the reference. If they want something different, it is not a biosimilar ....*
- *No, can't really answer this - the two sentences given above seem to be quite contradictory... So, my 'no' is no real 'no' but I had to tick one of the two options to proceed... Heterogeneity is part of the nature of biotech-manufacturing so it will not be avoided.*
- *I think that both are true; the challenge of course is that our understanding of the correlation of quality attributes and safety and efficacy profiles is very limited at present*
- *But, as noted above, both aspects need to be considered. This will present different sets of challenge on a product by product basis.*
- *In concept okay, but I am not sure how practical*

<table>
<tr>
<td></td>
<td>it is that one would develop an approvable process to mimic the Q-profile and then do process changes later unless there was a real motivation for the changes.
• Yes/no, mimic the product, but not its variations/heterogeneities, if these have no involvement in clinical performance.</td>
</tr>
<tr>
<td>Would you recommend to the biosimilar developer to initiate clinical trials (i.e. clinical comparability exercise), if the Q-profile of the reference biologic shifted and the QTPP of the biosimilar is based on the "old Q-profile" of the reference biologic.

1. Yes, as the risk of detecting a clinically significant difference due to changes of the Q-profile is a relatively low risk for the biosimilar developer.
2. Yes, as the "new Q-profile" of the reference biologic was approved by EMA and therefore should fall within similar safety and efficacy margins, as the "old Q-profile". Thus, it should not impact on the clinical comparability exercise.
3. Yes, as in the case where differences in clinical comparability exercises were detected in Q-profile of the biosimilar can be adjusted rapidly to follow the most recent Q-profile of the reference biologic.
4. No, as the Q-profile of the biosimilar should have prior clinical comparability</td>
<td>Implications in case where the reference biologic changes its quality profile, (question 1.2).

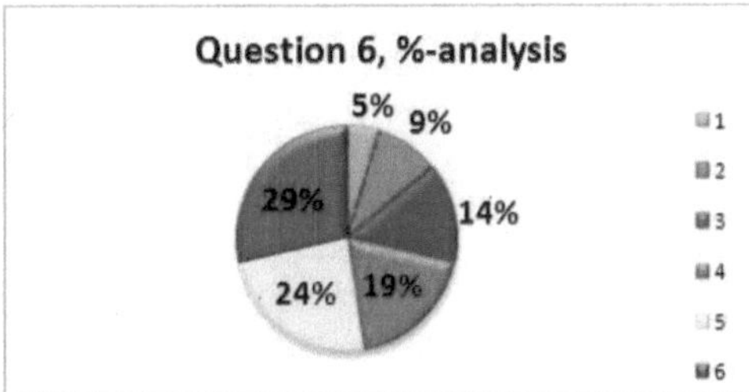

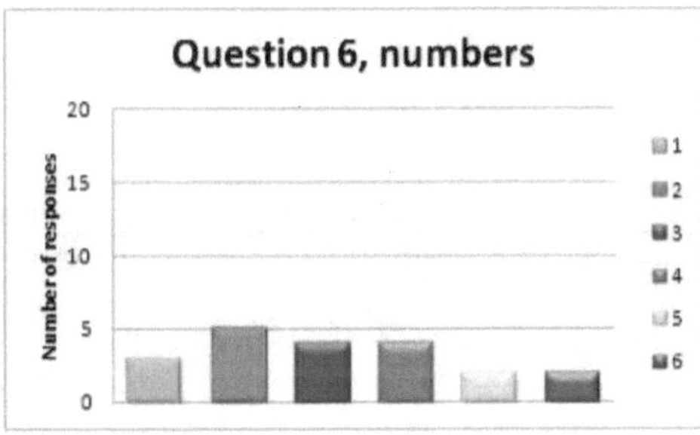

Note: The numbers in the legends of the two above figures correspond to the options given in the question in the coloum directly to the left.

Additional comments from participants:
• None of the choice above strictly apply. What regulators are learning, is that the reference in the field has a higher degree of variability than expected. As long as the BS is in the range of the reference profile, safe use can be predicted and</td>
</tr>
</table>

exercises and a very similar Q-profile compared to the reference biologic. This should result in a similar clinical outcome. 5. No, as differences during the comparability beyond the margin or tolerance would require additional investigation and the burden of those is deemed as not desirable. 6. No, as the risk of detecting a clinically significant difference due to changes in the Q-profile is a relatively high risk for the biosimilar developer.	*the trial should be able to start with low risk.* • *This is 'it depends' - basically depends on the rationale for the change/shift. If it was for a more pure or less immunogenic molecule then the old 'version' should not be developed and approved. If it is a shift that was not intended for safety or efficacy but as a effect of a process change made more for efficiency of manufacturing, then okay to develop and compare to either 'version' or be between them.*

2) The following findings were summarized from questionnaire 1.2.

Table: 17
Results from *"Questionnaire 1.2"*, refer to Appendix: 4, for more details.
The questions were sent to dedicated experts, refer to section: **3.3.1 Practical aspects of the questionnaires and critical considerations**.

Questions from questionnaire 1.2	**Results to research question**
Do you think that within one lot of a complex biologic (such as a mAb), it will always contain variations of mAb molecules (due to product or process related heterogeneity), which are slightly different to each other?	Research question 1.0 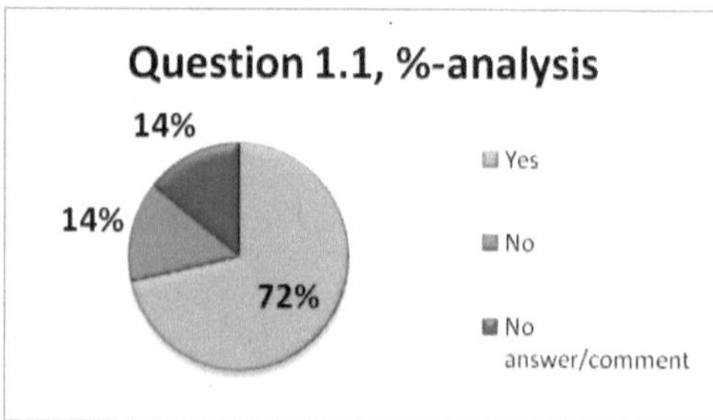

<table>
<tr>
<td></td>
<td>
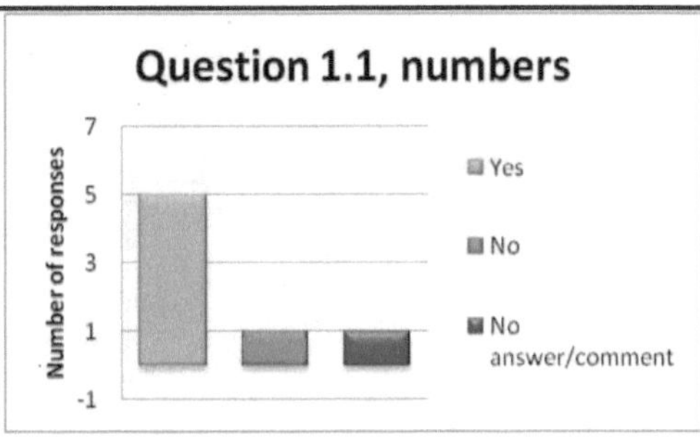

Additional comments from participants:

- *Some variability but within specified ranges*
- *batch to batch variability, the process is the product*
- *Absolutely - a variety of glycoforms, C terminal heterogeneity, deamidated, species etc. Even true to lesser extent for simple molecules such as somatropin – you will get deamidated species, etc*
- *This is not my expertise to answer.*
- *For a lot of complex biologic usually including a complex mix of large and complex molecules, which produced by glycosylation, deamidation, oxidation, amino acid modification and so on, all these variations caused the inherent characteristics of heterogeneity of complex biologic.*
</td>
</tr>
<tr>
<td>Do you think that the ratio of those slightly different (mAb)-molecules is unique for an individual lot?</td>
<td>
Research question 1.0

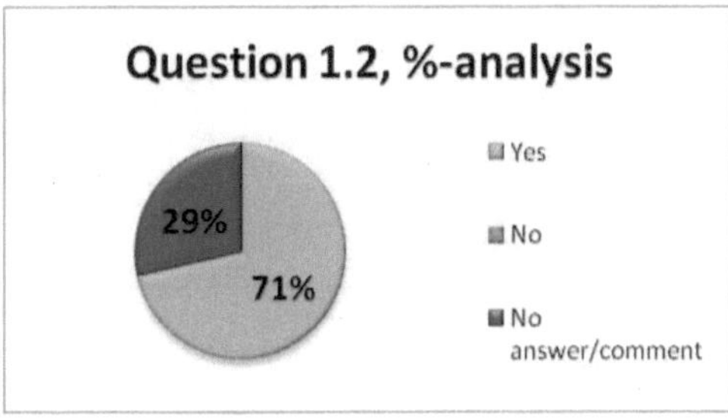

</td>
</tr>
</table>

<table>
<tr>
<td></td>
<td>
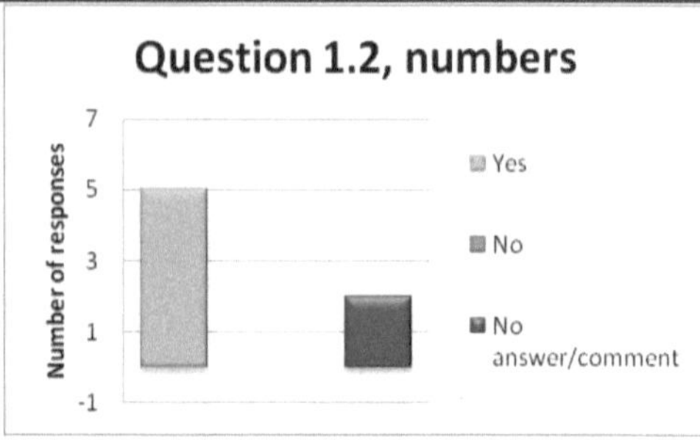

Additional comments from participants:

- *I presume you mean not reproduced by the next lot – there will inevitably be variability between lots in relative proportion of post translational forms – the question is how much – some of this is controlled within set limits.*
- *This is not my expertise to answer.*
- *For a complex biologic, the process will in a large extent predetermine the variation status of the products. Therefore, after the process finished, an individual lot of product will go into a relatively stable state, in which variation happened slowly.*
</td>
</tr>
<tr>
<td>Do you think that between multiple lots of a complex biologic (such as mAb), there will always be lot to lot variation of the Q-profile (due to product or process related variation and/or process changes)?</td>
<td>
Research question 1.0

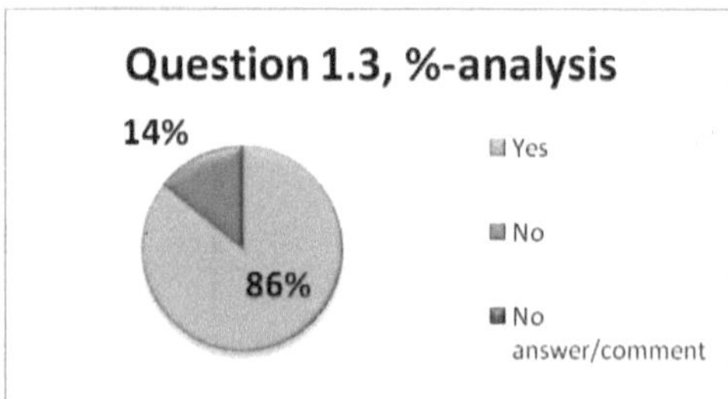

</td>
</tr>
</table>

<table>
<tr><td></td><td>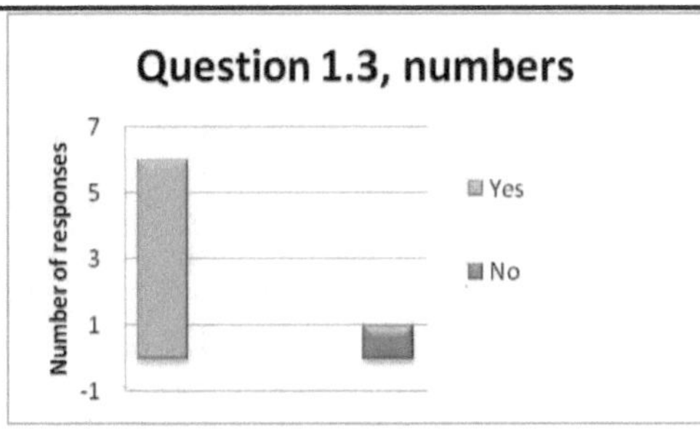

Additional comments from participants:
<ul>
<li>I presume you mean not reproduced by the next lot – there will inevitably be variability between lots in relative proportion of post translational forms – the question is how much – some of this is controlled within set limits.</li>
<li>Depends on selections of lots; several lots of DP does not necessarily correspond to several lots of DS</li>
<li>Even for non-complex products, the lot-to-lot variability exists. For complex biologics, there are even more factors that can contribute to variability. Therefore, it seems logical to assume that lot-to-lot variability is a given.</li>
<li>As mentioned above, a lot of complex biologic trends to form millions of slightly different molecules even in a fixed process. The key point is that how many variations will be emerged in an individual lot. Therefore, predetermined limits for different kinds of attributes based on understanding of the product are very important.</li>
</ul></td></tr>
<tr><td>Do you think that multiple lots of the biosimilar against multiple lots of the reference biologic need to be part of any comparability exercise?</td><td>Research question 1.0</td></tr>
</table>

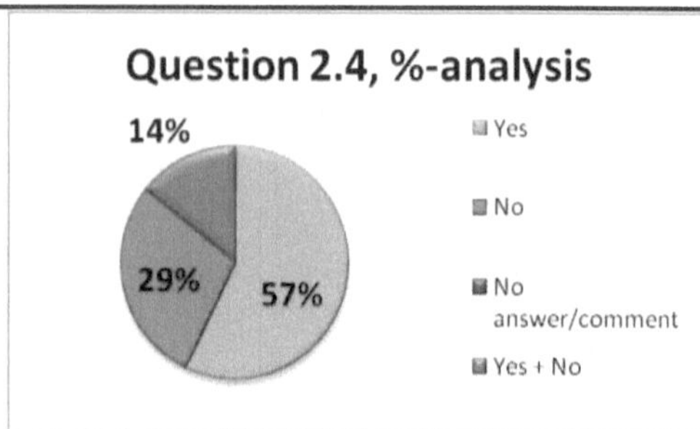

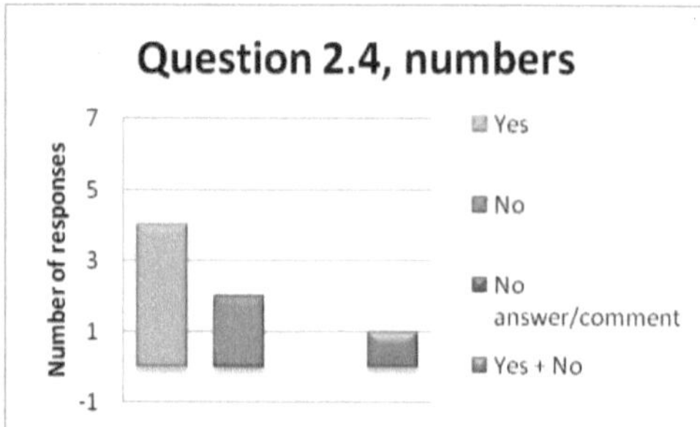

Additional comments from participants:

- *One needs to understand the range of variability of biosimilar as much as for reference product – three batches minimum preferably more*
- *Depends on the purpose of the comparability exercise. The demonstration of similarity would certainly need multiple lots of both, while demonstration of comparability after a process change would depend on the change introduced, and may not necessarily need comparison to the reference product (case by case).*
- *Yes, comparing multiple lots of both products will help to put results of certain assays with inherently large variability into perspective (for instance, cell-based assays).*
- *Since the safety and efficacy of the reference biologic had been proven by the clinical use. In the*

	PK/PD and the needed clinical comparability exercise, only one lot of the reference biologic might be enough used as a comparator for multiple lots of the biosimilar
Do you think that multiple lots of the reference biologic need to be analyzed, to understand the allowable limits for biosimilar manufacturing specification?	Research question 1.0 and 1.1 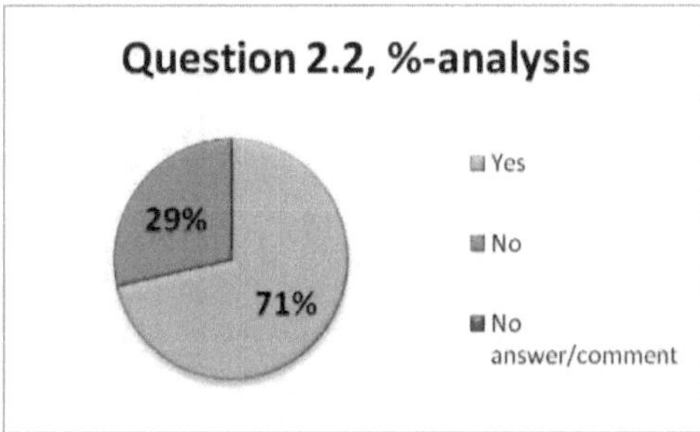 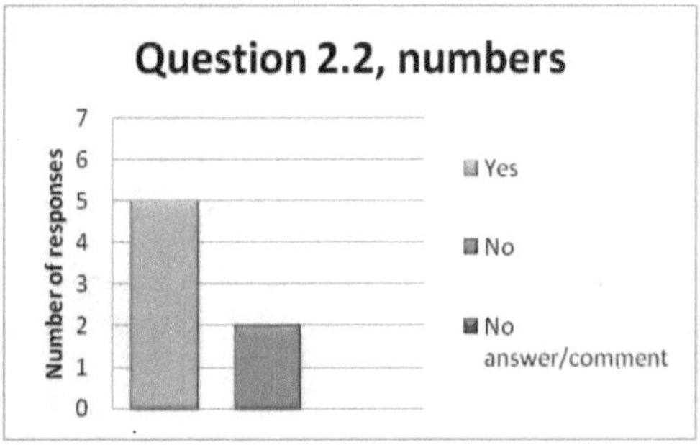 **Additional comments from participants:** • *Not if you define the critical comparibility criteria up front and justify the allowed ranges* • *In order to know the "range of similarity", lots should be analysed and extremes identified* • *Clearly one needs to understand the extent of lot to lot variation particulraly for critical parameters – also bear in mind that different lot numbers could still come from the same fermenation run* • *Target range and specification are two different*

<table>
<tr>
<td></td>
<td>
limits. The control of multiple lots can contribute to the setting of specification, but it is not sufficient alone. Specification should be set in accordance to ICHQ6B as for any biologics.
<ul>
<li>Although it is agreed that different batches should be tested, it must be stressed that testing multiple lots of the originator cannot provide the full scope of the quality profile of the originator product. (please see question above)</li>
<li>Multiple lots of the reference biologic are indispensable to understand the quality range of the originator, which will be the prerequisite for set the range for control of biosimilar products.</li>
</ul>
</td>
</tr>
<tr>
<td>What are the 5 major factors contributing to variation and heterogeneity of a biologic (such as mAb) in your opinion?
Rated by significance.</td>
<td>
Open question
<table>
<tr><th>Factor 1</th><th>Factor 2</th><th>Factor 3</th><th>Factor 4</th></tr>
<tr><td>Process Variability</td><td>Raw Materials</td><td>No answer given</td><td>No answe given</td></tr>
<tr><td>cell line drift</td><td>process optimization</td><td>formulation</td><td>storage a handling</td></tr>
<tr><td>Glycoforms (by far most important)</td><td>Aggregates</td><td>Split forms i.e. missing 1 light chain</td><td>Deamidat and oxidation</td></tr>
<tr><td>Manufacturing process and its development (including strain selection)</td><td>No answer given</td><td>No answer given</td><td>No answe given</td></tr>
<tr><td>No answer given</td><td>No answer given</td><td>No answer given</td><td>No answe given</td></tr>
<tr><td>glycosylation</td><td>deamidation</td><td>oxidation</td><td>terminal modificat</td></tr>
<tr><td>different expension system</td><td>different formulation</td><td>different packaging</td><td>No answe given</td></tr>
</table>
</td>
</tr>
</table>

<table>
<tr>
<td>Should a biosimilar developer focus on mimicking the Q-profile of the reference biologic including all its variability,
<u>Or</u>
focus on homogeneity, for example, of the major isoform?
Please elaborate your thoughts and comment!</td>
<td>Research question 1.0 and 1.1

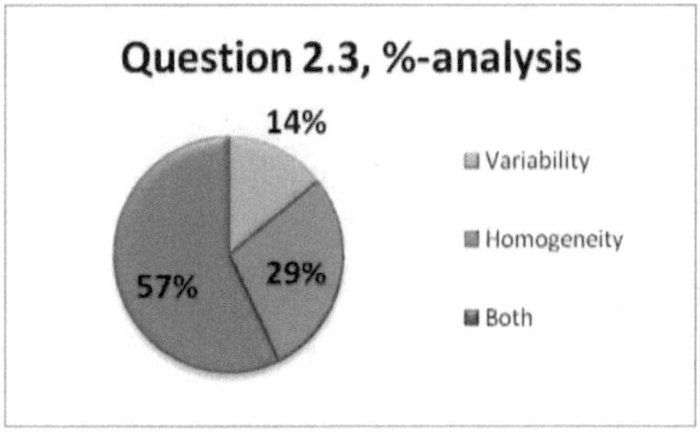

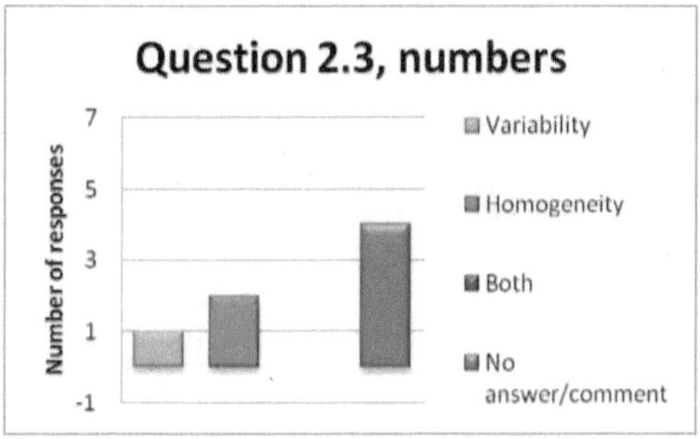

Additional comments from participants:

- *Focus on the major isoform(s) and critical criteria for QSE*
- *Should be within the range of the innovator.*
- *It helps to be as close as possible in that it reduces the need to justify differences but if justified any difference can be acceptable. Differences that impact on biological activity particularly relevant activity will be more problematic. But differences in c-terminal lysine can have a major impact on ion exchange profile but are insignificant from a clinical perspective.*
- *Mimicking as close as possible the quality profile would certainly make registration easier as it may decrease the level of "regulatory burden" (i.e.*
</td>
</tr>
</table>

<table>
<tr>
<td></td>
<td>extensive justification and demonstration that the differences observed to not significantly impact the safety and efficacy profile).
o An improve profile may have an impact on the safety and efficacy profile (e.g. need of dose reduction, different PK…). In such case, biosimilar pathway may not be the most suitable legal pathway to register the product.
• The strategy for biosimilar development differs from compound to compound and from biosimilar company to biosimilar company.
• I think a biosimilar developer should focus on homogeneity the major isoform of the critical quality attributes but not focus on mimicking the Q-profile of the reference biologic including all its variability. Firstly, although the reference biologic was established by years of good clinical use, its profile of variability may be not the best due to the historical limitation of the process level. Secondly, a biosimilar developer needs to use the state-of-the-art analytical methods to study the Q-profile of the product. New variants and isoforms might be detected by the advanced analytical methods.</td>
</tr>
<tr>
<td>How much would you agree with the following sentence?

"The quality profile of a biosimilar should fall within the limits of the quality profile of the innovator biologic, because one would expect that safety and efficacy will likely be similar too."</td>
<td>Research question 1.1

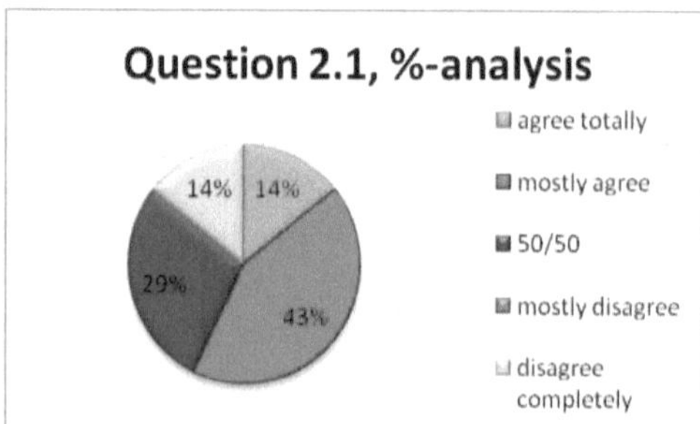
</td>
</tr>
</table>

	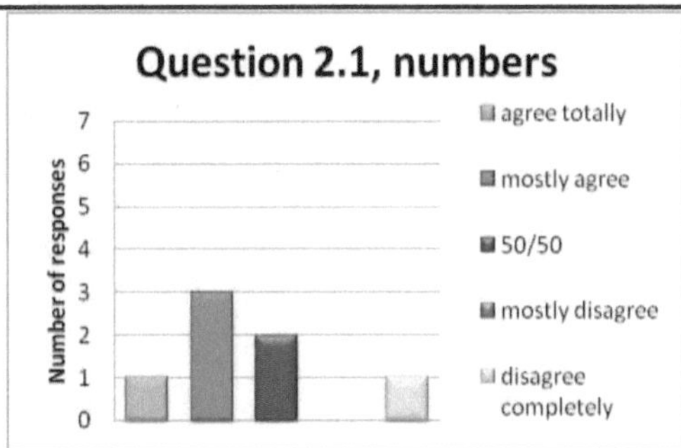 **Additional comments from participants:** • *I agree with the chmp guidance which states differences are acceptable if justified. Many variants are of no biological consequence eg c terminal lysine is removed within minutes of injection* • *It is expected that the range of variability of the biosimilar quality profile falls within the range of its reference product, however this is not possible for all quality attributes (eg process related impurities), and there is always a certain degree of uncertainty dependent on the complexity of the structure which could impact the safety and efficacy profile, thus this assessment generally cannot only rely on the quality profile.* • *In theory: yes, this is agreed. However, <u>in practice</u>, using the quality profile of the innovator will put the biosimilar is at a disadvantage, with tighter specifications than the originator. See below for explanation:* ○ *The quality profile + specifications of the originator product are usually not completely known for the biosimilar manufacturer. Although sometimes specifications (for some of the quality attributes) are known (which is not the same as the actual range in the quality profile of*

<table>
<tr>
<td></td>
<td>the batches that are released!), the biosimilar manufacturer has to build up knowledge of the innovator product by buying and testing different batches. Therefore, the biosimilar manufacturer can only know for a number of different innovator batches what the range in quality profile is. As the innovator batches that are released will all be within their specification, and only very few batches will be 'on the limit', the quality profile (and associated specifications) of the biosimilar will be similar or even more narrow than the originator. Biosimilar manufacturers can never obtain all information on the range of the quality profile of an innovator product, unless such information will be shared freely (for instance, in the EPAR of the originator).
• I agree with the sentence when we talk about the critical quality attributes, however, if the differences emerged in the attributes not related to clinical relevance, the out of limits might be accepted.
• The quality profile of the ref is not fixed either therefore; the statement cannot be totally agreed.</td>
</tr>
<tr>
<td>Would you recommend to the biosimilar developer to initiate clinical trials (i.e. clinical comparability exercise), if the Q-profile of the reference biologic shifted and the QTPP of the biosimilar is based on the "old Q-profile" of the reference biologic.

1) Yes, as the risk of detecting a clinically significant difference due to changes of the Q-profile is relatively low risk for the biosimilar developer.
2) Yes, as the "new Q-profile" of the</td>
<td>Research question 1.2
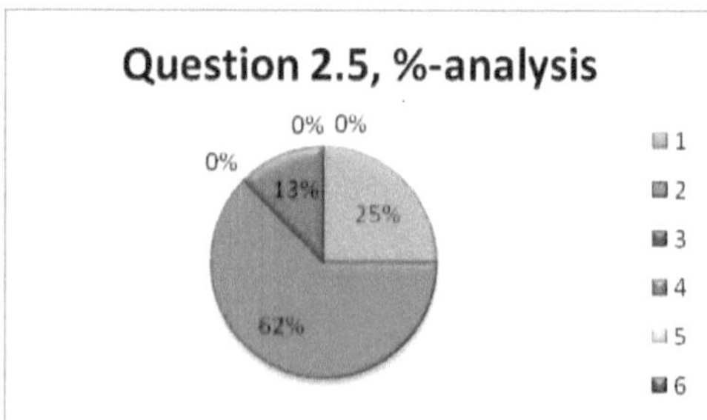
</td>
</tr>
</table>

reference biologic was approved by EMA and therefore should fall within similar safety and efficacy margins, as the "old Q-profile". Thus it should not impact of the clinical comparability exercise.

3) Yes, as in the case where differences in clinical comparability exercise were detected Q-profile of the biosimilar can be adjusted rapidly to follow the most recent Q-profile of the reference biologic.

4) No, as the Q-profile of the biosimilar should have prior the clinical comparability exercise a very similar Q-profile compared to the reference biologic. This should result in a similar clinical outcome.

5) No, as differences during the comparability beyond the margin or tolerance would require additional investigation and the burden of those is deemed as not desirable.

6) No, as the risk of detecting a clinically significant difference due to changes of the Q-profile is relatively high risk for the biosimilar developer

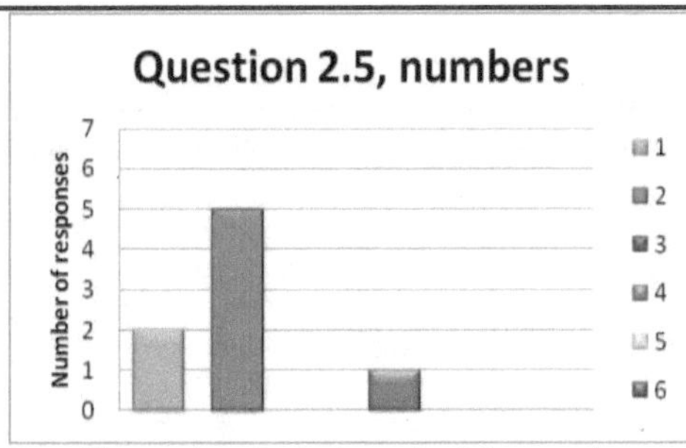

Note: The numbers in the legends of the two above figures correspond to the options given in the question in the coloum directly to the left.

Additional comments from participants:

- *Main discussion point at EMA: if the Q-range of innovator shifts over time, the total range for the biosimilar should also be "expanded", not "moved", as the initial Q-profile was accepted as is the new one. To be proven however in comparability file.*
- *All the comments above apply – differences are acceptable if justified*
- *For option 2, I don't fully agree with the last part of the sentence, as it would still be expected that the clinical trial confirms that there is no clinical impact.*
- *None of the above. The answer of this question depends on how the 'new' reference product compares to the biosimilar product. Obviously, in case of clear shifts in the quality profile, it is highly recommended to re-do the comparability exercise (bio-analytical and critical cell-based assays) with new reference batches.*
- *Due to the safety and efficacy of the reference biologic which approved based on the "old Q-profile" were verified by years of clinical use, QTPP of the biosimilar based on the "old Q-profile" of the reference biologic was enough to initiate clinical trials.*

4.4 Discussion

1.0 "What are the implications of heterogeneity and variation through the life cycle of the biosimilar and the reference biologic, from a European perspective?"

The underlying assumptions of this study about heterogeneity and variation were confirmed during the literature review, refer to section: 2.2 Life cycle in relation to heterogeneity and variation that cites factors which contribute/cause the heterogeneity and variation, complemented by the answers to questionnaire 1.2 (Appendix: 4) in section: 4.3.3 Results from the survey research method (Table:17) with the answers to the questions "What are the 5 major factors contributing to variation and heterogeneity of a biologic (such as mAb) in your opinion?". The acquired answers confirm and complement each other. Also, the notion of heterogeneity and variation of biologics is agreed with by experts, refer to Table:17 and the question "Do you think that between multiple lots of a complex biologic (such as mAb), there will always be lot to lot variation of the Q-profile (due to product or process related variation and/or process changes)?". Together with the EMAs view that biologics are complex mixtures, refer to section: 2.3.2 Guideline on comparability of medicinal products containing biotechnology derived proteins as active substance: Quality issues EMEA/CPMP/BWP/3207/00 (50) and the guide EMEA/CHMP/BWP/49348/2005 which subdivides variability in:

i) Product variability

ii) Process related variability.

Those two variability types form what is called the "Quality Profile" of the biosimilar and all biologics, refer also to Appendix: 2.

Those considerations remain largely undisputed; with small exceptions refer to Appendix: 5 for the answers of the EMA when stating:

> *"The "quality profile" is these two sorts of variability PLUS the usual stand-alone physicochemical and biological characterisation PLUS a comprehensive comparability exercise. As such, the "quality programme" for a biosimilar is more extensive than that of a stand-alone."*

and in Appendix: 2 where there are responses to these comments, as they were not clear, with the following:

> *3.1 First, I would like to ask EMA to kindly clarify some comments on the quality profile and to highlight if these points were incorporated into this guide as announced...*

It is not likely that EMA will contradict/put in question their guidance documents, as part of the anticipated answer to Appendix: 2. Therefore, also taking into account that no additional comments were received from questionnaire 1.1 and questionnaire 1.2 together with the rich available literature related to heterogeneity and variation, this is consideration to be solid and acknowledged by all stakeholders.

Based on the above, the next logical step is to assess the implications of that heterogeneity and variation.

Controversies are present already with the mention by EMA of "highly purified protein" in a context where it appears that only highly pure proteins are suitable to serve as a reference for a biosimilar, refer to section: 4.3.1.1 Screening of regulatory literature (EMA) for research question series 1. This view contains two complex problems: firstly there is neither a definition available for highly purified proteins nor any standard that allows the assessment of a protein to determine if it is highly purified. Even if a protein is found to be highly pure, this does not mean in any way that in fact it is so. The emphasis is that there are no identical protein molecules derived from a process, as already described in section: 2.2 Life cycle in relation to heterogeneity and variation. Section: 4.3.1.2 Screening of non regulatory literature and articles for research question series 1 describes this controversy with the expert statement that:

> *"....more likely to be applied to highly purified products, which can be thoroughly characterized (e.g., biotech-derived medicinal products), than to other types of biologics. The implicit assumption is, therefore, that the pathway does not apply to more poorly purified biologics that are even more complex and difficult to characterize than a highly purified recombinant biologic..."*

This statement confirms the view that the biosimilar pathway it subject to flexible interpretation on what is regarded as "highly pure" or "pure enough". Further improvements in analytical techniques may show that innovator biologics that were regarded years ago as being highly pure are now more heterogeneous. Therefore, in the case where differences are not detected, this reveals only that the employed analytical methods are not sensitive enough, referring to section: 4.3.1.2 Screening of non regulatory literature and articles for research question series 1.

The majority of experts agree that within one lot of a complex biologic there will always be variations of mAb molecules (due to product or process related heterogeneity), which are slightly different to each other and that the ratio of those slightly different bio-molecules is unique for an individual lot. As a consequence, between multiple lots of a complex biologic (such as mAb), there will always be lot to lot variation of the Q-profile (due to product or process related variation and/or process changes), refer also to section: 4.3.3 Results from the survey research method.

The other major issue is that this view limits the biosimilar approval pathway itself, in a manner that appears to be ambiguous, as introducing additional constrains outside the original scope as outlined by Article 10(4) of Directive 2001/83/EC states:

> *"Where a biological medicinal product which is similar to a reference biological product does not meet the conditions in the definition of generic medicinal products, owing to, in*

particular, differences relating to raw materials or differences in manufacturing processes of the biological medicinal product and the reference biological medicinal product..."

Assuming that the reference biologic is not a "highly purified protein" what is the justification for not allowing to be used for the biosimilar pathway? In theory the biosimilar pathway should be open to all biologics. The mentioning of highly purified proteins takes likely practical consideration into account, associated with mimicking quality profiles that are very complex; however, this is a challenge for the biosimilar industry and their technical abilities and know how, but a regulatory constraint, which leads to limited the biosimilar pathway to only "highly purified protein".

Controversy two is about the sameness of the reference material that shall be used in the comparability exercise, refer to section: 4.3.1.1 Screening of regulatory literature (EMA) for research question series 1. Of course it is advisable to measure and assess the biosimilar to a stable reference i.e. the same material.

This is the likely intention/reason from the EMA by insisting on sameness. The other aspect which can be regarded/based on this, is to limited the reference material to the EFTA area, resulting in EMA having control and jurisdiction over the innovator product used as a reference, which gives the advantage of knowing firsthand from the reference manufacturer, if major changes to the production process occurred or are planned, together with the power to approve or reject those.

However the issue that poses this requirement goes much deeper to the principles and nature of biologics. Results from section: 4.3.3 Results from the survey research method to the question: "Do you think that multiple lots of the reference biologic need to be analysed, to understand the allowable limits for biosimilar manufacturing specification?", with its mostly positive answer, refer to section: 4.3.1.1 Screening of regulatory literature (EMA) for research question series 1 and section: 4.3.1.2 Screening of non regulatory literature and articles for research question series 1 that applies also to this context.

The reference material, in line with this study is dynamic in its quality profile; thus it cannot be, strictly speaking, the same reference material. The reference biologic in its quality profile is inevitably moving over time and each lot is unique, with a unique purity and impurity profile. Surprisingly the results for the literature research, section: 4.3.1.2 Screening of non regulatory literature and articles for research question series 1, did not reveal any publication on the topic. Therefore, it could possibly be concluded that with "same reference material" it is only referred to as the material commercially available within the EU and that serves as a basis for the reference for the biosimilar, but neglecting the other interpretation the same reference material actually has.

If this means that batches/lots shall have a similar degree of heterogeneity and variation when compared to each other, this accommodates the dynamic of the quality profiles (research question 1.0) and gives guidance in line with research question 1.1.

If consistency makes reference to a lesser degree of heterogeneity and variation, then this can be interpreted in the context of "highly purified proteins" as a more pure product with a less dynamic quality profile. This would result in biologic, which has potentially a different safety or efficacy profile compared to the reference biologic, thus it is not a biosimilar anymore.

The main controversy related to the results from the literature research, section: 4.3.1 Results from the literature, as it is stated, is that the dynamic of the quality profile may well occur, however it is not described what to do in cases where it occurs.

> *"It is acknowledged that the manufacturing process of the reference medicinal product may evolve. Through its lifecycle, and May lead to detectable differences in some quality attributes. Such events could occur during the development of a biosimilar medicinal product and may result in a development according to a QTPP which is no longer fully representative of the reference medicinal product available on the market.*
>
> *The ranges identified before and after the observed shift in quality profile could normally be used to support the comparability exercise at the quality level, as either range is representative of the reference medicinal product. Quality attribute values which are outside the range(s) of variability measured in the different profiles of the reference medicinal product should be appropriately justified with regard to their potential impact on safety and efficacy.*
>
> *It should also be noted that there is no regulatory requirement for re-demonstration of biosimilarity once the Marketing Authorisation is granted."*

In fact, the quality profiles of both biologics, the reference and the biosimilar, are likely to be changing over time. From a biosimilar point of view this will commence once the biosimilar development process begins. The implications of this situation are that the dynamic development process of the biosimilar needs to be synchronised with the dynamic evolution of the reference biologic, and this is a potential major problem within the biosimilar concept.

A biosimilar developer may well be able to understand the heterogeneity and variation of the reference biologic very well; however, the risk always remains that those established ranges for the reference biologic shift at an unfortunate time point in the biosimilar development program. Potentially the most unfortunate time point is when clinical trials were being conducted. The EMA has made it clear that once the M.A. is granted for the biosimilar it is expected that both products may vary independently. This has unsolved implications for interchangeability as hinted by Declerck et al., as the dynamic of the quality profiles of the reference biologic and the biosimilar might lead over time to differences in the safety and efficacy profiles. As discussed in chapter 7, the EMA will have then the burden to ensure that changes to safety and efficacy of both the reference biologic and the biosimilar do not occur, due to process changes. A task which is in difficult to control with a certain level of residual risk, thus prone to failure over time (in my view).

However, also prior to the M.A. being issued for the biosimilar, the situation is not clear.

There are a couple of possible scenarios and the current state of the biosimilar regulatory approach does not precisely describe what to do. Thus "unclearness" and the resulting flexibility makes sense from the EMA point of view, as it allows a flexible regulatory approach to a flexible situation, that provides from EMAs point of view the most appropriate response. However, the counter side of that is this flexibility can be viewed as a risk by the industry and therefore be a discouraging obstacle for biosimilar development. This is a major constraint, but protection should be possible for the biosimilar industry, related to development timeframes and associated costs.

As the requirements imposed by EMA remain depending on each of the scenarios being "flexible", thus the costs and success of a biosimilar development program is difficult to estimate. Therefore, EMA should provide biosimilar developers with an as clear as possible guide about regulatory requirements, that takes the dynamic of both biologics into account. The submitted questions will trigger a response by EMA and any response is new, as the draft concept itself is not yet approved; thus, these help to fundamentally highlight to EMA foreseeable issues, and help EMA to tackle those in the future. As an extension this allows access to affordable and safe biological medicines. Therefore the relevance of this study fulfils any academic standards.

Taken together, there is no solution proposed by EMA for research question 1.0, besides the mentioning that <u>once the M.A. is obtained the biosimilar and the reference have their own lifecycle</u>. Refer to the proposed solution in Chapter 7.0 Integrated discussion.

1.1 Taking into account the new amendment of EMA/CHMP/BWP/617111/2010, with emphasis on the life-cycle of the Biosimilar, does the Biosimilar need to achieve a quality profile which falls within the quality profile of the reference biologic? Under which circumstances could this approach be deviated from?

After taking the notion of the dynamic of the quality profiles (reference and biosimilar) into account, there is a close link on how to tackle this dynamic in a comprehensive development strategy for biosimilars.

Experts agree that in order to capture the quality profile of a biologic that multiple lots of the biologic need to be analyzed, to understand the allowable limits for biosimilar manufacturing specification/range, refer to section: 4.3.3 Results from the survey research method. This forms what is called the quality profile of a biologic. The EMA mirrors this view in its most recently proposed guide "Guideline on similar biological medicinal products containing biotechnology-derived proteins as active substance: quality issues (revision 1)", as it is proposed to characterize the reference biologic based on combined scientific knowledge gained from the reference biologic. This results from an analysis of multiple lots of the reference biologic (at different stages of its shelf-life) in line with most expert views that, "multiple lots of the reference biologic need to be part of any comparability exercise", refer to section: **4.3.3 Results from the survey research method**.

Further, next to the uniqueness of each individual lot mentioned above one also needs to take into account the uniqueness of the manufacturing process, which is most likely different between the reference biologic and the biosimilar and as outlined for the biologics context, in Annex II of EudraLex Volume 4 (75), the process defines the product.

Both are subject to dynamic changes, as discussed in this study. Therefore, the question of the development aim or strategy of a biosimilar is of vital interest for the biosimilar developer. As part of section: **4.3.2 Results from the questions and comments submitted to EMA** the following proposition was submitted to the EMA, refer to **Appendix: 2**:

> *"The initial or cardinal biosimilar development aim is to mimic the quality profile of the reference biologic drug substance including its variation/heterogeneity (on a lot basis) where possible."*
>
> *"The better the correlation of the molecular characteristic (including process understanding) of the biosimilar drug substance with quality attributes is understood, the more the biosimilar developer can shift the focus (of the development strategy) from mimicking the Q-profile of the reference biologic, including its variability/heterogeneity, towards manufacturing a more consistent/pure biosimilar drug substance; and therefore, eliminating some of the heterogeneity of the reference biologic, as long as the safety and efficacy remain comparable within the margins of equivalence."*

A similar proposal where the emphasis and the degree on similarity is explained as the similarity of the reference biologic, as being similar to itself, which makes the biosimilar and the reference biologic in their clinical performance confounded.

> *"The Biosimilar requires to such an extend similarity to the Reference Medicinal Product, as the Reference Medicinal Product is similar to itself, when compared on a batch to batch basis, as the optimum. Deviation of this approach is generally acceptable, but requires justification and possibly additional clinical evaluation in order to assess associated risks."*
>
> *"Further, to guide the industry and avoid unnecessary costs and disappointments, it should be stated that:*
>
> *For non clinical setting it is advisable that the "Quality Profile", (refer to [(1)]) of the Biosimilar is within the levels of variation observed for the "Quality Profile" of the Reference Medicinal Product.*

Thoroughly determination of the "Quality Profile" of the Reference Medicinal Product can be used to set the margins of tolerance (for variation) for the Biosimilar "Quality Profile". Deviation of this approach is generally acceptable, but requires justification and possibly additional clinical evaluation in order to assess associated risks".

Based on section: **4.3.2 Results from the questions and comments submitted to EMA**, as a result of questionnaires 1.1 and 1.2 most experts agree with that statement. This is presumably linked to the potential clinical behavior, as in the case where the quality profile of the reference biologic is overwhelmingly characterized and linked to quality attributes, potential "safety and efficacy will likely be similar too" as agreed or mostly agreed upon by experts according to section: **4.3.3 Results from the survey research method**. The above proposed development aims where partially taken over/incorporated by EMA, refer to section: **4.3.2 Results from the questions and comments submitted to EMA**.

"The target acceptance criteria used in the comparability exercise should be justified. Quantitative limits should be established, where possible. The relevance of these limits should be discussed, taking into account the number of reference medicinal product lots tested, the quality attribute investigated and the test method used. These limits should not be wider than the range of variability of the representative reference medicinal product batches, unless otherwise justified".

However next to the ranges that are one aspect in answering question 1.1, the other aspect is what those ranges should incorporate:

1) Should a biosimilar developer focus on mimicking the Q-profile of the reference biologic including all its variability ("The quality profile of a biosimilar should fall within the limits of the quality profile of the innovator biologic, because one would expect that safety and efficacy will likely be similar too")

Or

2) Focus on homogeneity, for example, of the major isoform?

This question remains unclear from the EMA as it is referred to in section: **4.3.1 Results from the literature**:

- Scientific knowledge gained from the characterization of the reference biologic.
- The requirement to justify potential differences.

The proposed new EMA guide "Guideline on similar biological medicinal products containing biotechnology-derived proteins as active substance: quality issues (revision 1)", can be understood to prompt the developer to mimic the quality-profile of the reference biologic as well as possible, as this increases the chances of having similar clinical effects as well.

This very reasonable approach is diminished in its positive impact on the development strategy by leaving it open to interpretation. The view that this point is open to interpretation is also confirmed by section: 4.3.3 Results from the survey research method table 16 and table 17, where the participants of both questionnaires left a wealth of often opposite comments for this research question.

Taken together the data from this study suggests that it is required to mimic the purities and impurities of the reference biologic and to justify why the biosimilar is less impure i.e. has a more pure quality profile with less heterogeneity and variation (for purities and impurities).

1.2 What is proposed if the reference biologic changes its quality profile during the biosimilar development program?

Section: 4.3.1 Results from the literature revealed that there are practical limitations to how much reference biologic material can be used for the establishment of the QTTP for the biosimilar, refer to:

> *"Stockpiling the RMP for a long study is not practical. This example demonstrates just one of the difficulties faced by biosimilar trial .... Manufacturers of branded products are likely to use sophisticated defensive tactics, including the development of complex biopharmaceuticals, to maintain share."* (25)

The innovator industry on the other hand is free to improve and change their process for a given biologic at any stage. EMA actually encourages process changes as this is part of the desired process improvements in line with ICH Q8 R2. Such change might well represent an improvement for the biosimilar as the purity might be increased or the variation might be reduced; however, this represents a change of the QTPP of the biosimilar. Suddenly the biosimilar has a different Q-profile compared to the reference biologic and the clinical trial outcome is very uncertain. Three different possibilities might occur:

1. The Q-profile shift of the biosimilar is of no clinical relevance when conducting a clinical head to head comparison with the biosimilar that was developed based on the old profile. The reason for that could be that the shift was within the range of the old quality profile.
2. The Q-profile of the reference biologic results in, for example, an improved safety profile, as the impurities were reduced.
3. The Q-profile results in a different efficacy, due to better purification techniques a specific isoform can be selected, which improves the efficacy of the reference biologic. This could be a side effect when improving the safety profile, for example, and could pass unnoticed.

Even if all those scenarios may occur, the majority of experts agree (refer to section 4.3.3 Results from the survey research method tables 16 and 17) that even if the quality profile of the reference biologic may have shifted during the biosimilar development phase, clinical trials of both biologics with unequal quality profiles can still be conducted based on the assumption that only changes in the quality profile of the reference biologic will have no (major) impact on safety and efficacy, refer to the view that "...*the "new Q-profile" of the reference biologic was approved by EMA and therefore should fall within similar safety and efficacy margins, as the "old Q-profile". Thus, it should not impact on the clinical comparability exercise..."* with the biosimilar. However this view is put in question by the same participants as most agree that it would be advisable during the development of the biosimilar to mimic the quality profile of the reference biologic, as the resulting safety and efficacy will be likely similar too. However related to research question 1.2, the findings from section: 4.3.1 Results from the literature are probably the most significant. Dr. Kowid Ho (who kindly supported this study), stipulation that the quality profile over the entire life cycle of the reference biologic may well be dynamic, with the view that:

> *"Changes in quality attributes may be acceptable if it is demonstrated that there are no negative impact on safety and efficacy. Drift of quality attributes, the quality profile could occur over time,*
> *Challenges for the biosimilar developer when defining the target variability of reference product attributes could occur.*
>
> *Dr. Ho mentioned for acceptable changes, that depending on the quality attribute and their potential impact on safety and efficacy, could support and justify differences observed at the time of submission.*
>
> *Dr. Ho also raised the question of whether biosimilars can really remain similar throughout their entire lifecycle as they will also have their own and independent lifecycle once they receive marketing authorization".* (9)

From the practical experiences gained with EMA, as reported by Declerck et al., that head to head trials with the finally approved biosimilar and the reference biologic were not carried out, refer to section: 4.3.1 Results from the literature. This indicated that Dr.Ho`s view as a member of the EMA was already practiced by EMA in the past.

The EMA wrote that glycosylation is expected to be different between the biosimilar and the reference biologic, for example; however, changes need to be justified and investigated from the point of view of the impact on safety and efficacy, refer to section: **4.3.1.1 Screening of regulatory literature (EMA) for research question series 1** Table 15 "Guideline on similar biological medicinal products containing biotechnology-derived proteins as active substance: quality issues (revision 1)". As this guide is new and only in a draft form it is advisable to clarify the concept from EMA's point of view.

Taken together, sudden changes in the Q-profile of the reference biologic can jeopardize the comparability exercise of the biosimilar in the case where those changes have an impact on the safety or efficacy.

Chapter 5.0: What should be the scope of trials?

Research questions:

2.0 What should be the scope of trials for biosimilars?

2.1 How extensive do biosimilar trials need to be in order to show comparability?

2.2 Under what circumstances does one need to test multiple lots of the biosimilar vs. the reference medicinal product in trials?

5.1 Introduction

It is well known that clinical trials are the major cost driver during the development of innovator medicinal products. Therefore, whether or not to extrapolate clinical trials is the largest dispute between the biosimilar and innovator industries: as it represents a significant cost hurdle. The research question for this chapter is therefore: What should be the scope of trials for biosimilars?

Clinical trials are usually conducted in various sites around the globe and they are often carried out by clinical research organisations on behalf of the innovator industry.

Clinical trial programs are not standardised:
1) They are customised based on previous experience with the entity, for example.
2) Or depending on the aim of the trial (for example, comparator studies, which are relevant for biosimilars, as part of the comparability exercise).

Based on "**statistically relevant sample size**" (statistical power[60] is a measure of the accuracy of the results of the trials, which is based on the chosen confidence intervals[61] and chosen probability[62] settings), usually a hypothesis is sought to be confirmed or dismissed through the trial.

---

[60] Statistical power
The ability of a study to demonstrate an association or causal relationship between two variables, given that an association exists. Bandolier - http://www.medicine.ox.ac.uk/bandolier/booth/glossary/statpow.html

[61] Confidence intervals
Intervals are commonly chosen such that the parameter falls within a 95 or 99 percent probability, called the confidence coefficient. Britannica Academic Edition.

[62] Probability
A measure that quantifies the uncertainty associated with an event. Bandolier - http://www.medicine.ox.ac.uk/bandolier/booth/glossary/Prob.html

In addition, the experimental setting of the trial, such as inclusion of a **"placebo"** (an entity that has a similar appearance to the medicinal product but does not contain any active substance) or **"blinding"** of the trial (i.e. single blinding: the patient volunteer does not know if they get a placebo or the medicinal product, double blinding: the research organisation and the patient do not if placebo or medicinal product is administrated) or "**randomization"** (each study subject is randomly selected/planned to be given the placebo or the medicinal product) that all influence the size of trials.

Classically, in clinical trials for innovator medicines the scope has included patients and healthy volunteers, usually in the order of a couple of thousand healthy volunteers and patients in total. As a rule of thumb the following 5 phases could be considered as applicable (for innovator medicinal products):

Clinical trial phase 0:

Number of involved patients/volunteers	Aim of the trial
Very few volunteers ca. 1-5	Initial safety trials on a new medicine. Severe and imminent safety issues (adverse events) are investigated.

Clinical trial phase I:

Number of involved patients/volunteers	Aim of the trial
Few volunteers ca. 20-120	Screening for safety issues, such as adverse events. Assessing pharmacokinetic profiles. Determination of the dose range used for follow on testing for single and for multiple doses.

Clinical trial phase II a:

Number of involved patients/volunteers	Aim of the trial
Moderate number of patients ca. 60-150	Assessment of efficacy (and safety) in selected populations of patients with the indication the medicinal product will be prescribe for.

Clinical trial phase II b:

Number of involved patients/volunteers	Aim of the trial
Moderate number of patients ca. 100-350	Rigorous determination (very controlled trials) of efficacy (and safety) in patients with the disease (pivotal trials[63]) with distinct clinical endpoints[64].

[63] Pivotal trial:

Clinical trial phase III:

Number of involved patients/volunteers	Aim of the trial
Large number of patients over 1000 for innovator medicinal products	Large trials for the efficacy (and safety) in relevant patient populations to show additional patient benefit.

Clinical trial phase IV:

Number of involved patients/volunteers	Aim of the trial
Large number of patients, from hundreds to thousands. This may be over longer timeframe (years).	For example, additional groups of potential patients can be investigated, such as children, different genders, pregnant woman etc. Various new formulations and dosages as well as new indications may be tested. Screening for adverse reactions or for the contradiction of those may be carried out.

A biosimilar however, is required to show similarity to the reference biologic, but not patient benefit per se, as this has been shown by the reference biologic. This notion adopted by EMA has clear implications on the scope of clinical trials for biosimilars. Therefore, classical testing as described above in clinical trials with 5 phases does not apply.

The exact scope of biosimilar trials is not described by EMA. The research questions were selected for this study due to the financial implications and regulatory vagueness.

5.1.1 Why is the dynamic of the Q-profile for biologics of relevance in clinical trials for biosimilars?

This question is interlinked with the main research question in Chapter 4.0: Main research question, as the heterogeneity and variation of the biosimilar and the reference biologic may have an impact on the clinical trials for the biosimilar. The quality profile divided into the purity profile (isoform ratios) and impurity profile (due to different expression system and different process) may have an impact on the success of the clinical trials for biosimilars and thus they impact on their scope as it is required to show comparability between both.

Typically a Phase 3 study which presents the data required by a regulatory agency to decide whether or not to approve a drug. A pivotal study will generally be well-controlled, randomized, of adequate size, and whenever possible, double-blind.
http://www.lillytrials.com/docs/terminology.html#P

[64] Clinical endpoint:
Refers to occurrence of a disease, symptom, sign or laboratory abnormality that constitutes one of the target outcomes of the trial.
http://en.wikipedia.org/wiki/Clinical_endpoint

As the quality profile of the reference biologic is dynamic, it is possible that during the comparability exercise when clinical trials where conducted the quality profile of the reference biologic has already changed and the safety and efficacy profile will be found to be dissimilar.

Also linked to the research questions:

- **How extensive do biosimilar trials need to be in order to show comparability?**
- **Under what circumstances does one need to test multiple lots of the biosimilar vs. the reference medicinal product in trials?**

For the question of how many batches of the reference biologic require testing against how many batches of the biosimilar, it is needed to show that the safety and efficacy profiles are similar in clinical trials. This fundamental question relates back to the idea, that the possibly different isoform profiles have a different impact on efficacy, whereas different impurity profiles have a different impact on safety. This is true when comparing "the reference biologic" with "the biosimilar". However, one needs to bear in mind that "the product" meaning an identical product does not really exist within the biologic context. The product varies, through its quality profiles and possible implications on its safety and efficacy profiles.

Therefore, the situation could arise, when assessing clinical data as part of the comparability exercise, where the biosimilar developer finds themselves in a position of having evidence that the safety or efficacy profile of the reference biologic has shifted.

This is a scenario that is not covered by any publication, to the author's knowledge. Thus, this can be viewed as an original idea developed during this study and further analysed during the course of this study.

5.1.2 Meta analysis for biosimilar clinical trials?

Within the general (for innovator medicinal products) clinical trial context, meta analysis for innovator biologics is common.

Meta analysis as a statistical technique in which the findings from various independent studies are combined with the intention of extracting, through this combination, new findings or to confirm existing ones. Usually randomised trials are the basis of meta-analysis, where the validity of the meta-analysis depends on the quality of the systematic review on which it is based. The quality of the systematic review and the statistical quality of the underplaying studies are critical for combinability of clinical trial studies. Often treatment benefit, i.e. efficacy, is investigated.

Good meta-analysis aims for complete coverage of all relevant studies, looks for the presence of heterogeneity, and explores the robustness of the main findings using sensitivity analysis.

The term meta-analysis was formally introduced in 1976 by the physiologist Gene V. Glass; however, the principles of meta-analysis as an analysis of primary findings and data (within the clinical trial context) date back to the beginning of the 19th century. Meta analysis is typically used in the clinical trials context when data and simple analysis requires further clarification by meta analysis. Thus, this is not the typical situation within the biosimilar context where margins of equivalence are acceptable. As patient benefit is not subject to the biosimilar trials, but comparability to the reference biologic as such, the term meta analysis is new to the biosimilar context.

One could argue that biosimilars should require a reduced scope of trials making meta analysis unnecessary, reciprocal to the idea of section: 5.1.1 Why is the dynamic of the Q-profile for biologics of relevance in clinical trials for biosimilars? (assessing the idea of multiple batches within clinical trials for biosimilars). Assuming head to head trials are conducted the answer from each trail should be clear. The design of the study can result in a clear answer.

<u>Thus, meta analysis should not routinely be required to clarify the clinical comparability exercise.</u>

As a potential outcome or confirmation of the ideas presented with in sections: 5.1 Introduction and 5.1.1 Why is the dynamic of the Q-profile for biologics of relevance in clinical trials for biosimilars?, meta-analysis for clinical trials, meta analysis would be typically carried out if a patient benefit cannot be proved by stand alone trials for innovator medicinal products, but not for biosimilars as comparability is required.

<u>The question is if and how does meta analysis apply to biosimilar clinical trials? If it does, could this be an indication that multiple batches needs to be analysed, taking into account the heterogeneity and variation of the biosimilar and possibly the reference biologic</u>.

Also, the amount of data that might be required to conduct meaningful meta analysis may not be available within the biosimilarity context. It is assumed that the study design to show successfully clinical comparability is limited.

Heterogeneity of data for meta analysis:
A major concern about meta-analyses is the extent to which it mixes studies that are different in kind (heterogeneity). The key difficulty lies in deciding which sets of studies are 'combinable' (76). That requires a careful and rational based selection of the data that will contribute to the meta-analysis.

5.2 Experimental procedure (methods and materials) employed

5.2.1. Literature research

Publications on biosimilars and meta analysis, in particular, were searched in MEDLINE (Pubmed) and by exploiting the public search engine Google Scholar. Key search words were combinations of the key words biosimilar and meta analysis. Searches were limited to full-text English-language articles. The websites from the European Medicines Agency (EMA) were also consulted.

Publications were also searched for in Gabi-online. Key search words were combinations of the words biosimilar and clinical trial. Searches were limited to full-text English-language articles.

The websites from the European Medicines Agency (EMA) were also systematically consulted and biosimilar guidance documents were selected and analyzed, refer to section 3.3.1 Practical aspects of the questionnaires and critical considerations for details.

5.2.2. Questions and comments submitted to EMA

Questions and comments were submitted to EMA utilising the form and contact details they provided, refer to Appendix: 1 Questions and comments submitted to the EMA and Appendix: 2 Questions and comments submitted to the EMA.

- The questions and comments for the Guideline on similar biological medicinal products containing monoclonal antibodies Draft, were submitted on the 30.05.2011, Appendix: 1 Questions and comments submitted to the EMA and Appendix: 2 Questions and comments submitted to the EMA and EMA published answers to those on the 30.05.2012, available under Overview of comments received on 'Guideline on similar biological medicinal products containing monoclonal antibodies' (EMA/CHMP/BMWP/403543/2010).

Table: 18
Extracts from questions and comments submitted to the EMA. Refer to Appendix: 1 for more details.
The questions and comments for the *"Guideline on similar biological medicinal products containing monoclonal antibodies"* were submitted on the 30.05.11 and the EMA provided responses to those questions and comments, refer to *"Overview of comments received on 'Guideline on similar biological medicinal products containing monoclonal antibodies"* (69).

Question and comments submitted to EMA extracted from Appendix: 1	Notes and clarification in relation to research question
For the clinical aspects, the same point as raised in [(1)] is valid.	Suitability of

<table>
<tr>
<td>
1.) Assuming the Biosimilar needs to be as safe and effective as the Reference Medicinal Product as indicated in lines: 209 to 212 and that the clinical efficacy of the Biosimilar is required to be similar to the Reference Medicinal Product.

2.) The PK studies require to be similar compared to the Reference Medicinal Product. Referring to lines: 239 and 240

3.) Also taking lines 254 to 256 into account,

clearly the safety and efficacy comparability exercise is subject to tolerance margins.

However, this margin of tolerance is not clear and would require further elucidation. What is the justification for a "generally acceptable" margin with regard to Biosimilars?

Taking into consideration the following points:

1.) The problems associated with populations for trials for example in a cancer setting where a sufficiently homogenous population might be rare and healthy volunteers cannot be used for ethical reasons either.

2.) Cases where the "Quality Profile" of the Reference Medicinal Product already exceeds an efficacy margin of 80-125% when compared on a batch to batch basis in the clinical trial to itself.

There is also additional clarification required, from the innovator industry's point of view, mainly on the following points:

1.) Taking the costs associated with clinical trials into account and the need to reduce healthcare costs, is it required to assess multiple batches of the Reference Medicinal Product when comparing it with the Biosimilar in a clinical trial setting, as a general question?

2.) As indicated in lines 209 to 212, the expectation is that safety and efficacy are "comparable".

A) Does this mean the acceptable tolerance margin for the Biosimilar in a clinical efficacy trial is required to be within the variation of endpoints of multiple batches of the Reference Medicinal Product and then the conventional equivalence margin of 80%-125% for the Biosimilar is applicable?
</td>
<td>the comment or question with regard to tackling the research question, can be better assessed when reading the whole context available in the appendix</td>
</tr>
</table>

Or B) Does this mean the acceptable tolerance margin for the Biosimilar in a clinical efficacy trial is based on one batch the Reference Medicinal Product and then the conventional equivalence margin of 80-125% for the Biosimilar is applicable?	

Refer to Appendix: 1 Questions and comments submitted to the EMA and Appendix: 2 Questions and comments submitted to the EMA for more details.

The questions and comments for the Guideline on similar biological medicinal products containing biotechnology-derived proteins as active substance: quality issues (revision 1) were submitted on the 28.11.2012 and the EMA has not published answers to them so far.

Table: 19
Extracts from questions and comments submitted to the EMA for "Guideline on similar biological medicinal products containing biotechnology-derived proteins as active substance: quality issues (revision 1 Draft". Refer to Appendix: 2, for more details. The questions and comments were submitted on the 28.11.2012 and the EMA has not published answers to these yet.

Questions and comments submitted to EMA, extract version	**Notes and clarification in relation to research question**
2.1 It is clear that some of the Q-attributes can only be partially predicted and tested by the methodologies mentioned in this guide (due to the predictive nature of those tests), see blow: *Various assays have been established in the past years that allow for more in-depth characterization of complex proteins, both on a physicochemical and a functional level, e.g. with potency assays, and there is experience in the assessment of minor quality differences due to changes in manufacturing processes for monoclonal antibodies. However, it may at the current stage of knowledge be difficult to interpret the relevance of minor quality differences in the physicochemical and biological characterization when comparing a biosimilar mAb to a reference mAb.*	Suitability of the comment or question with regard to tackling the research question, can be better assessed when reading the whole context available in the appendix.

<table>
<tr><td>

EMA/CHMP/BMWP/403543/2010

EMA therefore acknowledges the need to confirm those predictions in clinical trials. As QTPP terminology was consciously chosen by EMA for this guide. This should be incorporated more in this guide by EMA, by making the link with guidance documents such as *"Guideline on similar biological medicinal products containing monoclonal antibodies – non-clinical and clinical issues"* (*EMA/CHMP/BMWP/403543/2010*).

(III) therefore EMA should state in section 4 that:

"...at later development stages the established Q-attributes require confirmation in clinical trials and may be amended/completed, as the trial outcome cannot be foreseen. This could result in adjustment of the QTPP."

</td><td></td></tr>
<tr><td>

3.2.1 Scenario 1:

EMA permitted a process change for the reference biologic, which was considered acceptable. When compared head to head with the biosimilar in a clinical trial setting this change resulted in a measurable impact/difference (beyond the margins of equivalence).

The Q-profile change of the reference biologic shifted, whereas the QTPP of the biosimilar was based on the pre-change Q-profile of the reference biologic. The post-change reference biologic had to be used for clinical trial comparability exercises due to normal shelf-life limitations.

The biosimilar was shown to possess a better safety and efficacy profile compared to the reference biologic.

(Example mentioned verbally in London during last meeting with EMA about: "Guideline on similar biological medicinal products containing monoclonal antibodies – non-clinical and clinical issues".)

- What is the consequence of that for the EMA?
- What is the consequence of that for the reference product?
- What is the consequence of that for the biosimilar?

3.2.2 Scenario 2:

EMA permitted a process change for the reference biologic, which was considered acceptable. When compared head to head with the biosimilar in a clinical trial setting this change resulted in a measurable impact/difference

</td><td>Suitability of the comment or question with regard to tackling the research question, can be better assessed when reading the whole context available in the appendix.</td></tr>
</table>

(beyond the margins of equivalence). The Q-profile change of the reference biologic shifted, whereas the QTPP of the biosimilar was based on the pre-change Q-profile of the reference biologic. The post-change reference biologic had to be used for clinical trial comparability exercise due to normal shelf-life limitations. The reference was shown to possess a better safety and efficacy profile • Is the comparability exercise considered failed? • Is refining of the QTPP required and some re-development?	

Refer to Appendix: 1 Questions and comments submitted to the EMA and Appendix: 2 Questions and comments submitted to the EMA for more details.

5.2.3 Survey research

A series of questions related to the research question was made public as a survey in selected forums as part of questionnaire 1.1 and was submitted as part of questionnaire 1.2 to selected stakeholders, which included regulatory agencies, the innovator industry, the generics industry and consultants as well as other independent experts.

The following question was relevant to the research question investigated, within questionnaire 1.1, refer to Appendix: 3 Questionnaire 1.1 for details.

Table: 20
Extracts from *"Questionnaire 1.1"*, refer to Appendix: 3, for more details.
The questions were published online in a dedicated forum, refer to section: 3.3.1 Practical aspects of the questionnaires and critical considerations.

Question from survey monkey questionnaire 1.1	**Notes and relation to research question**
Do you test multiple lots of the reference biologic against multiple lots of the biosimilar during the clinical efficacy exercise?	Research question 2.0 and 2.1

The following questions were relevant to the research question investigated, within questionnaire 1.2; refer to Appendix: 4 Questionnaire 1.2 for details.

Table: 21
Extracts from *"Questionnaire 1.2"*, refer to Appendix: 4, for more details.
The questions were sent to dedicated experts, refer to section: **3.3.1 Practical aspects of the questionnaires and critical considerations**.

Questions from questionnaire 1.2	Notes and relation to research question
Do you think that for biosimilar comparability exercise the classical phase I, phase II and phase III approach is required? Please give your rational for or against that.	Research question 2.0 and 2.1
Would it make sense to test multiple lots of the reference biologic against multiple lots of the biosimilar during the clinical PK/PD studies?	Research question 2.0 and 2.1
Please comment if you would combine PK, PD studies (step 1) with clinical efficacy studies (step 2)	Research question 2.0 and 2.1
Do you think that almost always the steps 2 (clinical efficacy) of the comparability exercise is required?	Research question 2.0 and 2.1
Do you need to conduct this clinical efficacy exercise as a head to head comparison?	Research question 2.0 and 2.1
Do you test multiple lots of the reference biologic against multiple lots of the biosimilar during the clinical efficacy exercise?	Research question 2.2

5.3 Results for the series 2 research questions

5.3.1 Results from the literature

- **Regulatory literature was screened as part of chapter 2.3**

The follow regulatory literature topics impact on the research questions as summarised in table form. For the detailed text passage, please refer to section: 2.3 Screening the above presented literature related to current biosimilar regulation with regard to the research questions and the following, which contains comments from the author and further clarification.

Table: 22
Extracts from section 2.3. series. Highlighting topics and findings from the analysis of regulatory documentation with regard to the research question. For the detailed text passage, please refer to section: 2.3. series and the following, which contains comments and further clarification.

Summary of regulatory literature review, impacting research question series 2 / Notes related to research question	**Refer to section**
"The requirements to demonstrate safety and efficacy of similar biological medicinal products have to comply with the data requirements laid down in Annex I to Directive 2001/83/EC".	2.3.1 Guideline on similar biological medicinal products CHMP/437/04 (49)
Clinical trials and comparability exercise depend on: The 3rd chapter of the guide mentions that the process is different and comparing to public standards is not sufficient. Thus, an extensive comparability exercise is required. The extent of pre-clinical and/or clinical testing needs to be considered with regard to the complexity of the biopharmaceutical itself, as well being based on possible differences in the reference biologic itself. From the point of view of this study, those differences require a thorough investigation. The following factors should be taken into account for a comparability exercise: 1) The complexity of the molecules. 2) The type of changes in the process. 3) Their impact on quality, safety and efficacy.	2.3.2 Guideline on comparability of medicinal products containing biotechnology derived proteins as active substance: Quality issues EMEA/CPMP/BWP/3207/00 (50)
Immunogenicity: There is also a guide for the comparability exercise with regard to clinical and non-clinical issues available. Biosimilars were mentioned as well. The main concern was on safety and immunogenicity with regard to biosimilars.	2.3.3 Guideline on comparability of medicinal products containing biotechnology-derived proteins as active substance: Non-clinical and clinical issues EMEA/CPMP/3097/02/Final (51)

However there were no requirements outlined for efficacy. Input or discussion on clinical trial requirements impacting clinical trial size were not detected. Thus, this guide remains of a more abstract nature with regard to the research questions of this study.	
Clinical comparability exercise: *The design of comparative PK studies should not necessarily mimic that of the standard "clinical comparability" design (CHMP/EWP/QWP/1401/98), since similarity in terms of absorption/bioavailability is not the only parameter of interest. In fact, differences in elimination characteristics between products e.g. clearance and elimination half-life should be explored. The choice of the design for single dose studies, steady-state studies, or repeated determination of PK parameters should be justified by the applicant".* *"The pharmacodynamic (PD) markers should be selected on the basis of their relevance to demonstrate therapeutic efficacy of the product. The pharmacodynamic effect of the test and the reference medicinal products should be compared in a population where the possible differences can best be observed. The design and duration of the studies must be justified. Combined PK / PD studies may provide useful information on the relationship between exposure and effect".* **Conclusion/Comment:** • This is an indication that the clinical phase should be designed to detect differences in efficacy. • PK studies should focus on detecting possible differences. **PK/PD studies instead of clinical trials:** *"Normally comparative clinical trials are required for the demonstration of clinical comparability. In certain cases, however comparative PK/PD studies between the similar biological medicinal product and the reference medicinal product may be sufficient to demonstrate clinical comparability, provided that all*	2.3.7 Guideline on similar biological medicinal products containing biotechnology-derived proteins as active substance: Non-clinical and clinical issues EMEA/CHMP/BMWP/42832/2005 (55)

the following conditions are met...."

Conclusion/Comment:

- Comparability can also be shown based on PD/PK studies (under certain circumstances), which leavers the scope of phase III comparability studies for efficacy.

Clinical trials:

"Usually comparative clinical trials will be necessary to demonstrate clinical comparability between the similar biological and the reference medicinal product. Clinical comparability margins should be pre-specified and justified, primarily on clinical grounds. As for all clinical comparability trial designs, assay sensitivity (see ICH topic E10) has to be ensured. If a clinical comparability trial design is not feasible, other designs should be explored and their use discussed with the competent authorities. Data from pre-authorisation clinical studies normally are insufficient to identify all potential differences. Therefore, clinical safety of similar biological medicinal products must be monitored closely on an ongoing basis during the post-approval phase including continued benefit-risk assessment".

Conclusion/Comment:

- Differences due to insufficient similarity should be detected prior to the M.A., thus the EMA acknowledges the limitations of that concept and defines an increased monitoring interval to complement the similarity exercise.
- Clinical comparability trials are usually required, but this can be amended if not feasible.

Immunogenicity in safety trials:

"The periodicity and timing of sampling for testing of antibodies should be justified. In view of the unpredictability of the onset and incidence of immunogenicity, long term results of monitoring of antibodies at predetermined intervals will be required. In case of chronic administration, one-year

follow up data will be required pre-licensing". *"On the scope of trial also antibody testing is required not only for Abs against the biosimilar itself, but also for process related impurities".* **Conclusion/Comment:** • Long term (up to 1 year) sampling for antibodies against the biosimilar and impurities is required.	
Scope: A new concept is available for the scope of clinical trials: *"With regard to the phase I studies the guideline recommends the selection of relevant pharmacodynamic markers, but such markers are not always available and/or cannot be shown to reflect efficacy. On the other hand, when relevant surrogate markers do exist, the current guideline includes the possibility of using them (e.g. absolute neutrophil [65] count to assess the effect of granulocyte-colony stimulating factor) as the primary end points in the pivotal phase III studies. This aspect could be expanded to further elaborate on the underlying principles, potentially including examples from other therapeutic areas e.g. oncology. The current guideline does not mention any recommendations on the patient population in the pivotal phase III studies".* **Conclusion/Comment:** • Only phase I and phase III trials were mentioned. Phase II trials were not mentioned and may be not required. **Trial design:** *"The current guideline does not tackle the issue of an alternative to the equivalence design, which is currently expected for clinical studies to show sufficient similarity in efficacy (mentioned e.g. in the Draft Guideline on Similar Biological Medicinal Products Containing Monoclonal Antibodies).* *In the equivalence trial clinically meaningful*	2.3.8 Concept paper on the revision of the guideline on similar biological medicinal products containing biotechnology derived proteins as active substance: Non-clinical and clinical issues EMA/CHMP/BMWP/572828/2011 (56)

[65] Neutrophil:
It refers to a neutrophilic white blood cell. Oxford Online Dictionary.

differences including superior efficacy between the biosimilar and the reference product are excluded if the 95 % confidence interval of the treatment difference is fully contained within the pre-specified two-sided (upper and lower) comparability margins. *In the case of a non-inferiority study, only the lower margin is defined, i.e. the statistical analysis can only show that the product in not inferior to the reference product, but cannot exclude the possibility that the biosimilar product would be better than the reference product. The number of patients needed for an equivalence trial tends to be larger than that for the non-inferiority trial.* *Also the possibility of using a non-inferiority design in the pivotal phase III studies in certain cases will be considered."* **Conclusion/Comment:** • The trial design also impacts the size. In principle, there are two different designs: non inferiority design and the equivalence trials design. EMA does not exclude any of them from being carried out. **Phase III:** *"The BMWP will discuss, in case relevant markers are not available and similarity has been shown with selected PK parameters, whether the development of a biosimilar product can proceed to the pivotal phase III studies.* *With regard to the design of the phase III studies, the BMWP intends to update the guideline to include examples of surrogate endpoints which are being already recommended in various other biosimilar guidelines, e.g. in the Draft Guideline on Similar Biological Medicinal Products Containing Monoclonal Antibodies. The BMWP will discuss on how to further define considerations around acceptance of pharmacodynamics markers for demonstration of clinical similarity, and how to generate sufficient safety data in addition."* **Conclusion/Comment:** • Phase III is mentioned but phase II is not.	

Phase I: *"With regard to the phase I studies the guideline recommends the selection of relevant pharmacodynamic markers, but such markers are not always available and/or cannot be shown to reflect efficacy. Many biological products e.g. monoclonal antibodies, have several approved indications, and the principle emerges that the <u>most sensitive</u> one of them should be chosen to <u>show therapeutic similarity</u>".* **Conclusion/Comment:** • This means that it is most sensible for efficacy testing to be done in phase III trials.	

- **An additional review of non-EMA papers and articles with regard to the research question was carried out. Refer to section:** 3.1.2 Literature review.

5.3.1.1 Screening of non regulatory literature and articles for research question series 2

An additional review of non-EMA papers and articles with regard to the research question was carried out, as outlined in section: 3.1.2 Literature review.

Different scopes for different biosimilars
For the main question of interest in this chapter, the scope of trials presented here are typical examples of the situation described in the literature:

- For a biosimilar version of Filgrastin, a M.A. has been granted by EMA on the basis of clinical trials with approximately **900 patients**. (77)
- For the biosimilar Valtropin, (reference is Somatropin a recombinant human growth hormone (rhGH)), the pivotal study monitoring paediatric growth was a non-inferiority, randomised, 12-month double-blinded phase, followed by a 12-month non-blinded (open) extension phase of **149 prepubertal children 6-10 years old.** (74)
- For Omnitrop, the key study was a nine-month, open label, parallel design study in **89 pre-pubertal children aged 5-13 years old**, followed by an extension phase with a switch from comparator to Omnitrop. In both cases the adult indication was approved as well, by extrapolation from the pediatric study. (74)

Conclusion/Comment

This makes clear that the trial size varies, presumably influenced by many factors. The above leads to idea that the complexity of the biologic itself in combination with ethically justifiable patient protection reasons could influence the trial size for biosimilars.

Factors influencing trial scope

It has been reported that for mAbs clinical trials the guidelines for the design of clinical pharmacokinetics take numerous factors into account:

1. Long half-life of mAbs
2. Immunogenicity
3. Disease and patient characteristics
4. Pharmacokinetics of reference

The endpoints and pharmacodynamic markers should be selected and sensitive enough to detect possible differences between the biosimilar and the reference biologic, and should be evaluated to provide evidence of comparability. Further different doses response times may be needed to be tested. (26)

Conclusion/Comment

This result is for trials for biosimilars that depend on the scope of the biosimilar itself; however, it is difficult to comply with the mentioned test parameters within one single trial. Moreover, a series of successive trails might build one on another and be required to comply best with those requirements.

> *"Ultimately, clinical efficacy needs to be evaluated. Importantly the guiding principle is to demonstrate similarity of the biosimilar to the reference, and not to evaluate the patient benefit. Consequently, the selection of the patient population is mainly based on the need of homogeneity and sensitivity".* (26)

Examples of phase III trials for mAbs biosimilars

The current requirements for biosimilar trials seem not to be clear to all companies. For example, it is reported that Epirus Biopharmaceuticals demonstrated bioequivalence of a mAb to Infliximab in a single dose trial in early 2013. This single centre, double blind, randomized parallel group trial was preceded by a phase I trial with 84 healthy male volunteers. A controlled phase III trial is ongoing and results are expected in late 2013. (78)

However, the development of another mAb biosimilar appears to have been halted at the end of 2012. Samsung stopped its worldwide biosimilar trials program in some European countries, such as the Czech Republic and Spain, but is continuing in Hungary, Poland and the UK. (79)

TEVA also reported that phase III trials were halted due to different views on the requirement for trials with the EMA. (80)

Conclusion/Comment:
The most recent publications suggest that the scope of clinical trial requirements for highly glycosylated proteins such as mAbs is more complex than initially anticipated by biosimilar developers.

200 to 500 subjects as average
It was reported that from 2009 onwards, clinical studies for biosimilar trials would require 200 to 500 subjects, depending on the product class in question. This is a considerable reduction in the clinical scope for biosimilars compared to the innovator biologics. The exact number of required subjects depends on the biosimilar itself. (81)

Conclusion/Comment:
It is understood that this applies to mAbs biosimilars that are not yet approved. However, the link to the product class can be seen as a link to the complexity of the biosimilar and the degree of similarity that was achieved by the biosimilar developer, particularly for the quality profile.

Margins of tolerance

> *"The fact is that it is statistically impossible to show that two products are 100% equivalent; there has to be a margin of acceptable difference which needs to be defined prospectively".* (73)

Conclusion/Comment
The allowed margins of tolerance for the clinical trials are a significant boundary that limits the required scope of clinical trials. This assumes of course that the reference biologic meets the margins of tolerance itself, as if not, the margins of tolerance will not be wide enough for the biosimilar.

Endpoints and their margins of tolerance
Sensitive endpoints should be chosen in a way to have meaningful comparative criteria, able to measure the degree of similarity in the clinical efficacy of the biosimilar compared to the reference biologic. However, these sensitive endpoints were usual not employed when the reference biologic was approved, as surrogate endpoints were chosen.

> *"Sensitive endpoints are not surrogate endpoints, they are less ideal since small studies with large equivalence margins will yield less certainty around establishing biosimilarity."* (9)

Tolerance margins are predefined and justified based on statistical and clinical considerations. (9)

Conclusion/Comment
The margins of tolerance for the endpoints are controversial, since if wide margins are chosen, some treatment effects that could be shown by the reference biologic (especially for the less strong treatment effects compared to placebo trials) and have a great impact on patient health,

may results in a biosimilar that has reduced efficacy compared to the reference biologic and possibly no efficacy if it would be compared to an independent placebo trial (which is not required for biosimilars). In other words the endpoint should be sensitive (capable of detecting changes) but the clinical relevance of the endpoint is not important, as the focus of the biosimilarity exercise is comparability but not patient benefit, per se.
Sensitive endpoints are considered to require less patients and are quicker to carry out compared to the possibly very lengthy innovator trials for when assessing survival rates in cancer trials.

Placebo arm in clinical biosimilar trials
Treatment effects in clinical trials are usually evaluated vs. a placebo, called a "placebo arm" for innovator medicinal products. Thus it is assumed that the effect is constant with regard to time, patient population and location, etc. (9)

Conclusion/Comment
When conducting trials for biosimilars the focus is not on effects per se, the focus is on similarity of effects between the biosimilar and the reference biologic.

Combining safety aspects in efficacy trials
Dr. Schneider form the EMA suggested combining clinical safety studies with efficacy study. (9)

Conclusion/Comment
The possibility of combining multiple questions within one or a limited number of trials is very important. Thorough clinical planning beforehand for the sampling and subsequent analysis is required.

Phase I and phase III trials
As for any clinical development programme, monoclonal development will need to commence with a phase I study, following the CHMP "Guideline on similar biological medicinal products containing biotechnology-derived proteins as active substance: nonclinical and clinical issues" (EMEA/CPMP/42832/05) (56) that needs to be completed before advancing to confirmatory efficacy and safety studies. For biosimilars approved to date, these phase I studies have been performed in healthy volunteers, but with monoclonals this may not always be ethical or feasible because of the intense PD effect. Thus, the advocated phase I approach may not be appropriate or feasible for monoclonals and in some circumstances it may only be feasible to study the comparative PK in patients at steady state as part of the phase III programme. This is recognised in the draft biosimilar monoclonal guideline, which states that in such circumstances *"proper measures have to be pre-planned to ensure the statistical rigor and integrity of this trial"*. Nevertheless, the expectation will be that some data is available before commencement of phase III in order to mitigate against any risks to trial subjects. (73)

Conclusion/Comment
This is a clear direction based on the directive that phase I safety trials are required prior to proceeding to extended efficacy and safety trials (phase III).

The mode of action and its relevance for clinical trials

Dr. Kresse noted that antibody modes of action are complex and may involve contributions from multiple mechanisms, and that the in vivo net contribution of different modes of action described for one mAb are often not fully understood and may also be different in different indications. As a consequence of this multi-functionality, mAb characterization should include both Fab and Fc mediated functions unless there is justification to omit these studies. He emphasized that biosimilars must have the same amino acid sequence as the reference product, and that both reference and biosimilar mAb products will be micro-heterogeneous mixtures of a large number of post-translationally modified molecular species. The relevance of major variants on clinical efficacy and safety thus has to be established, and since the exact composition of the mixture cannot be reproduced using a different manufacturing process, comparative non-clinical and clinical data will always be necessary for biosimilar mAbs.

Differences between biosimilar and reference products can be accepted based on the level of understanding of clinical relevance. Bayesian statistics can help to optimize clinical trial sample size. (70)

Conclusion/Comment

The mode of action and the implications of all functional parts of the reference biologic, including possibly different modes of actions of different isoforms.

Current issues for phase III mAb trials

TEVA, a major biosimilar company has suspended its phase III trial for its biosimilar version (TL011) of Rituxan/MabThera (rituximab). TEVA is reviewing how best to meet the requirements' from FDA and EMA .

Samsung another biosimilar developer is required to recommence its clinical tests due to changes in safety requirements in Europe and the US. The company is reported to be in talks with FDA to reset the conditions for the SAIT101 clinical tests. (79)

Conclusion/Comment:

Major biosimilar developers struggle to assess correctly the requirements for biosimilar trials correctly. As a result of that, clinical trials have been required to be repeated.

5.3.1.2 Clinical trials conducted in the EU or USA

Data collected from EMA.

Gabionline and Pubmed were systematically reviewed and articles researched which indicated information on clinical trial size, refer to table 23.

Table: 23
Information related to phase III trials of biosimilars published on GABI Online (32) found using the keyword pairs: 'phase III/3 and biosimilar', 'biosimilar and meta analysis' and 'phase III/3 and meta analysis', with no limits on the search/publication dates.

Company	Product name	Reference product	Trial in EU or US	Phase III size	Time frame	Trial design	Trial objective	Number of groups	Subject information	Problems and issues with trial	Citation and date published
Boehringer Ingelheim	(BI 695500)	rituximab	Not indicated	Not indicated	September 2012 and is expected to be completed in April 2015	Randomized, double-blind, parallel arm, multiple dose, active comparator trial Consist of five groups of patients Two will receive the biosimilar and three will	Show the pharmacokinetic similarity and equivalence of efficacy of Boehringer Ingelheim's biosimilar (BI 695500) to Rituxan	Consist of five groups of patients	Patients with moderately to severely active rheumatoid arthritis, Adults aged 18 to 80 years	Not indicated	(82) 21/09/2012

						receive Rituxan All patients will receive two IV infusions of the drugs separated by a two-week wash-out period					
Finox Biotech	Bemfola	Gonal-f (follitropin alfa)	Not indicated	Not indicated	completed in June 2012	Pivotal phase III clinical trial Standard treatment duration of 10 to 16 days with a fixed dose of r-FSH	Not indicated	**Not indicated**	**Not indicated**	**Not indicated**	(83) 18/01/2013
Biocon	Regular human Insulin (Insugen R) and isophane human Insulin (Insugen N)	Actrapid and Insulatard	Not indicated	Nearly 300, T1DM patients	First part comoplete d The second part of the study is ongoing and is expected to be completed by 2013	The multi-centre, randomized study Conducted in nearly 300, T1DM patients Compared efficacy, safety and immunogenicity of Biocon's regular human	Met efficacy endpoint by demonstrating non-inferiority in HbA1c endpoint at six months	**Not indicated**	Nearly 300, T1DM patients	**Not indicated**	(84) 16/11/2012

					Final results expected in 2014	Insulin (Insugen R) and isophane human Insulin (Insugen N) against Novo Nordisk's originator products, Actrapid and Insulatard.					
Sandoz	Epoetin alfa (HX575)	Amgen's Epogen	US	Not indicated	Started recruiting patients in September 2012 Trial will be completed by September 2014	The trial is a multicentre, double-blind, randomized, active controlled, parallel-group study.	Evaluate the efficacy and safety of weekly subcutaneous injections of HX575 Compared against Epogen/Procrit (epoetin alfa) Patients with end stage renal disease Over a period of	Not indicated	Patients with end stage renal disease Over 18 years old	Not indicated	(85) 09/11/2012

							52 weeks				
Teva Pharmaceutical Industries	Version (TL011) of Roche's Rituxan/MabThera (rituximab)	Roche's Rituxan/MabThera (rituximab)	Europe	Not indicated	Started recruiting patients in May 2012 Now suspended	Multicentre, double-blind, randomised, active controlled, parallel-group study	Evaluate the efficacy, safety, tolerability and pharmacodynamic profiles of TL011 infusions Compared with MabThera (rituximab) Subjects with severe, active rheumatoid arthritis treated with methotrexate	**not indicated**	Subjects with severe, active rheumatoid arthritis treated with methotrexate Aged 18 to 80 years	According to Israeli newspaper *Haaretz* the company is 'considering how best to meet the requirements' from EMA and FDA.	(80) 12/10/2012
Celltrion	CT-P13	Infliximab	Not indicated	606	Not indictated	Randomised 1:1 to receive either CT-P13 or infliximab (3 mg/kg, 2-hour intravenous	Proved that CT-P13 and infliximab are equivalen	**Not indictaed**	606 patients with active rheumatoid arthritis	**Not indicated**	(86) 06/07/2012

						infusion per dose) plus methotrexate and folic acid Treatment period consisted of a dose-loading phase (weeks 0, 2, and 6) and a maintenance phase (weeks 14, 22, and 30)	t in terms of ACR20 in patients with rheumatoid arthritis				
Finox Biotech	Afolia	Gonal-F	Not indicated	410	Not indicated	Assessor-blinded, multicentre, phase III study 410 patients randomised using a 2:1 scheme to receive either the biosimilar (Afolia) or the reference product (Gonal-F)	Afolia demonstrated clinical and statistical equivalence to the reference product, Gonal-F	**Not indicated**	Not indicated	**Not indicted**	(87) 18/06/2012
Celltrion	CT-P13	Infliximab	Not indicated	617	During 2010 and 2011	617 patients were followed for approximately one year	Celltrion's biosimilar proved to be equivalent in terms	**Not indicated**	Patients with rheumatoid arthritis	**Not indicated**	(88) 17/02/2012

						Treatment of rheumatoid arthritis with Remicade or Celltrion's Infliximab biosimilar	of efficacy, safety and drug dynamics to the reference biological (Remicade).				
Hospira	Erythropoietin (EPO)	Amgen's Epogen	US	Will enrol approximately 1,000 patients	Results from the phase III programme are expected in 2013	not indicated	Compare safety and efficacy of Hospira's EPO with Amgen's Epogen Patients with anaemia associated with chronic renal (kidney) failure and chemotherapy	**Not indicated**	Approximately 1,000 patients On haemodialysis Already been treated with Epogen From more than 200 haemodialysis centres across the US	**Not indicated**	(89) 13/01/2013
Sandoz GmbH	Filgrastim (rhG-CSF) Zarzio	Neupogen (Amgen GmbH)	Not indicated	170	Not indicated	170 chemically naive, female, breast cancer patients	Not indicated	**Not indicated**	170 chemically naive, female, breast	**Not indicated**	(90) 04/06/2010

						Received Zarzio (Sandoz GmbH), at a dose of 300 mg or 480 mg administered via the SC route			cancer patients		

Table: 24
Information related to phase III trials of biosimilars published on Pubmed (31) found using the keyword pairs: 'phase III/3 and biosimilar', 'biosimilar and meta analysis' and 'phase III/3 and meta analysis', with no limits on the search/publication dates.

Company	Product name	Reference product	Trial in EU or US	Phase III size	Time frame	Trial design	Trial objective	Number of groups	Subject information	Problems and issues with trial	Citation and date published
Hospira	Filgrastim	Amgen filgrastim	EU	184	Not indicated	Randomized (2:1) to receive Hospira filgrastim or Amgen filgrastim After the end of chemotherapy Filgrastim (5 µg/kg/day) was administered	Hospira filgrastim and Amgen filgrastim are bioequivalent in efficacy with similar safety profiles	Not indicated	Breast cancer patients Suitable for treatment with doxorubicin and docetaxel in the neoadjuvant /adjuvant or first-line	Not indicated	(91) 2010

						under double-blind conditions Primary endpoint to demonstrate bioequivalence was duration of severe neutropenia (DSN) in cycle 1			metastatic Enrolled at 37 European centers		
Sandoz	New somatropin [Somatropin Powder]	A licensed reference rhGH preparation	EU	89 + 51	Not indicated	Not indicated	Study 1 – 89 children Compare efficacy and safety of the lyophilized formulation of the new somatropin [Somatropin Powder (Sandoz)] with a licensed reference rhGH preparati	2	Children	Not indicated	(92) 2007

							on and the liquid formulation of the new somatropin [Somatropin Solution (Sandoz)] Assess long-term efficacy and safety of this ready-to-use Somatropin Solution Study 2 - 51 children Efficacy and safety of Somatropin Powder and to confirm its low immunogenic potential				

							rhGH was given sc at a daily dose of 0.03 mg/kg				
Filgrastim	A new granulocyte colony-stimulating factor (G-CSF) biosimilar, XM02	Not indicated	Not indicated	608 patients (363 under XM02 and 245 under filgrastim)	Not indicated	608 patients (363 under XM02 and 245 under filgrastim) Meta-analysis 3 clinical studies	Compare a new granulocyte colony-stimulating factor (G-CSF) biosimilar, XM02, with filgrastim Terms of prophylactic effect on the development of febrile neutropenia (FN) during the first chemotherapy cycle	Not indicated	608 patients (363 under XM02 and 245 under filgrastim) Conducted with breast cancer, lung cancer, and non-Hodgkin's lymphoma patients	Not indicated	(93) 2009
Sandoz BioPharm	Recombi	Not indicated	EU	70	Not indicated	Naïve, prepubertal	Demonstrate the	Not indicated	Prepubertal Spanish	Not indicated	(94) 2011

aceuticals, Holzkirchen, Germany	nant human GH Omnitrope					Spanish children (n=70) with isolated GHD Treated with Omnitrope 0.03 mg/kg/day subcutaneously	safety and efficacy of long-term treatment with the recombinant human GH Omnitrope		children (n=70) with isolated GHD		
Not indicated	Epoetin zeta	Epoetin alfa	Not indicated	216	Not indicated	Open, non-controlled, multiple-dose, international, multicentre, phase III study 216 patients with solid tumours or non-myeloid haematological malignancies receiving chemotherapy and at risk of transfusion	Evaluate epoetin zeta, a biosimilarepoetin referenced to epoetin alfa, for the treatment of chemotherapy-induced anaemia in patients with cancer	Not indicated	216 patients with solid tumours or non-myeloid haematological malignancies	Not indicated	(95) 2009
Not indicated	G-CSF, XM02	Filgrastim	Not indicat	92	Not indicated	92 patients receiving	Investigate a new	Not	92 patients receiving	Not indiacted	(96) 2009

			ed			chemotherapy Randomised in cycle 1 to treatment with daily injections (subcutaneous 5 microg/kg/day) of XM02 (n = 63) or filgrastim (n = 29) Treatment at least 5 days and a maximum of 14 days	G-CSF, XM02, in comparison to filgrastim in terms of safety and efficacy in the prevention of chemotherapy-induced neutropenia in non-Hodgkin-lymphoma (NHL)	indicated	chemotherapy		
BioGeneriX AG	XM02	Filgrastim Neupogen	Not indicated	240	Not indicated	Multinational, multicenter, randomized, controlled phase-III study 240 patients were randomized to treatment with either XM02 (*n* _ 160) or Filgrastim (*n* _ 80) in the first chemotherapy	Show new granulocyte colony-stimulating factor, XM02, is as safe and effective as Neupogen Treatment of	Not indictaed	240 patients receiving platinumbased chemotherapy	Not indicated	(97) 2009

						cycle In the subsequent cycles, all patients received XM02	chemothe rapy-induced neutrope nia in patients with small cell or non-small cell lung cancer				

5.3.2 Results for the questions and comments submitted to EMA

Refer to Appendix: 1, Appendix: 2 and Appendix: 5 for details.

1) The questions and comments for the *"Guideline on similar biological medicinal products containing monoclonal antibodies"* were submitted on the 30.05.11 and the EMA provided responses to those questions and comments, refer to *"Overview of comments received on 'Guideline on similar biological medicinal products containing monoclonal antibodies"* (69), also represented in Appendix: 5.

- No particular answer was made from the EMA to the research questions in this section, refer to Appendix: 5.

2) Questions and comments for the *Guideline on similar biological medicinal products containing biotechnology-derived proteins as active substance: Quality issues (revision 1)* were submitted on the 28.11.2012. (A response from EMA is still outstanding as of April 2013.)

5.3.3 Results for the survey

The following findings are summarized from questionnaire 1.1. For more detail, refer to Appendix: 3 Questionnaire 1.1. In cases where a potential controversy was observed all comments from the participants were included in this study. Whereas, in the cases where an overwhelming majority was reached no further comments were included.

Table: 25
Results from *"Questionnaire 1.1"*, refer to Appendix: 3, for more details.
The questions were published online in a dedicated forum, refer to section: 3.3.1 Practical aspects of the questionnaires and critical considerations.

Questions from survey monkey questionnaire 1.1	**Responses from questionnaire**
Do you test multiple lots of the reference biologic against multiple lots of the biosimilar during the clinical efficacy exercise?	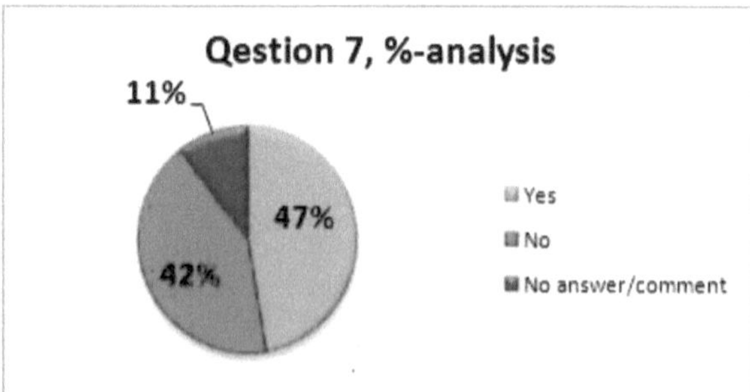

<table>
<tr>
<td></td>
<td>

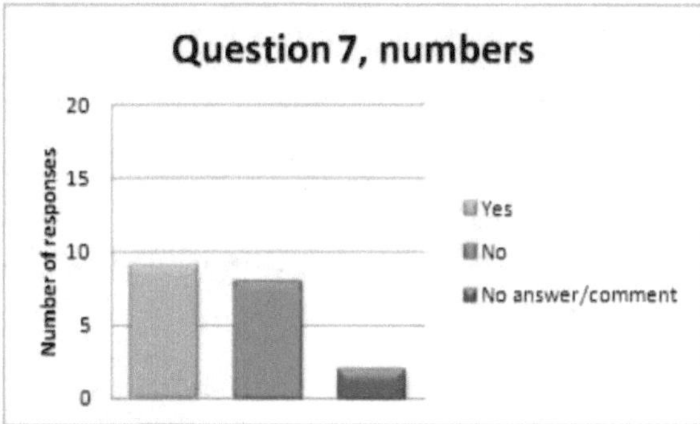

Additional comments from participants:

- *This would increase patient number and increase variables. I would not follow this approach*
- *can't answer this, I am not working in a pharma company but in an association*
- *The range of quality attributes of the reference product should be established before clinical trials for a biosimilar*
- *In practice though, for larger trials multiple lots end up being used.*
- *No, we do not test; we supply lots for testing*
- *Especially if doing a PK study followed by a larger clinical study then it is more than likely more than one lot will be used to complete both studies.*
- *It would be foolish and likely make trials useless to mix ref. product lots. In the U.S., FDA expects biosimilar applicants to only prove clinical biosimilarity to a single ref. product lot.*

</td>
</tr>
</table>

The following findings are summarized from questionnaire 1.2. For more detail, refer to Appendix: 4 Questionnaire 1.2. In cases where a potential controversy was observed all comments from the participants were included in this study. Whereas, in cases where an overwhelming majority was reached no further comments were included.

Table: 26
Results from *"Questionnaire 1.2"*, refer to Appendix: 4, for more details.
The questions were sent to dedicated experts, refer to section: 3.3.1 Practical aspects of the questionnaires and critical considerations.

<table>
<tr><th>Questions from questionnaire 1.2</th><th>Results to research question</th></tr>
<tr><td>What do you think that about meta analysis within the biosimilar context? Please indicate where you agree and comment.

1) Meta analysis should not be required, as the clinical comparability exercise should be clearly designed with a clear outcome.
2) Meta analysis should not be required, as the summary of data generated for the comparability exercise are not usually not sufficiently large to be used in a statistical meta analytical way.
3) Mentioning of Meta analysis by EMA may indicates that EMA foresees situations where the clinical comparability exercise is not clear based on head to head comparisons.
4) Mentioning of</td><td>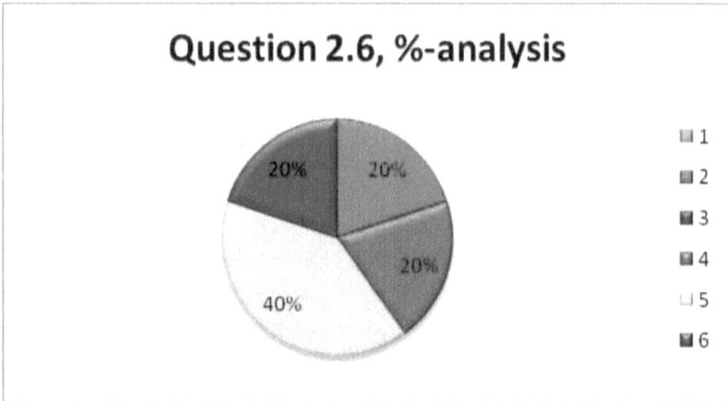

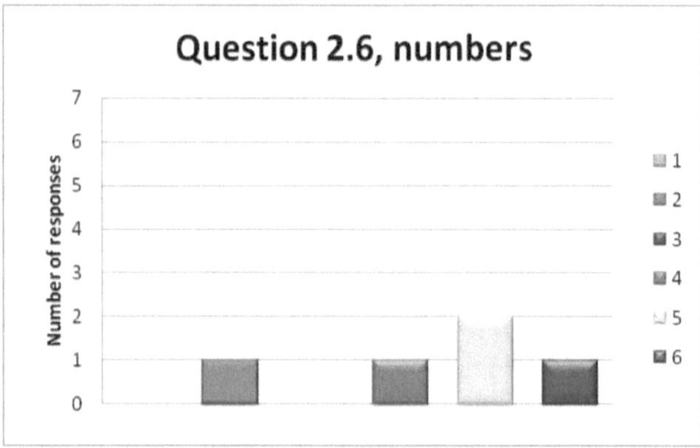

Note: The numbers in the legends of the two above figures correspond to the options given in the question in the column directly to the left.

Additional comments from participants:<ul><li>Meta analysis could be part of the post approval Risk management plan</li><li>Beyond my expertise</li><li>I think none of the above. In my opinion, the mentioning of meta-analyses refers to assess the effect of the innovator compound, which on itself serves the writing of the clinical trial protocol(s) that are aimed at identifying a difference</li></ul></td></tr>
</table>

Meta analysis by EMA may indicate that EMA foresees situations where the summary of clinical data are substantial enough to conduct Meta analysis. 5) Meta analysis is to welcome, as it allows to combine data and to re-assess the outcome. Thus more confidence in the clinical comparability exercise is generated. 6) Meta analysis is to welcome, as it is also be part of many innovator biologic clinical trial data analysis. Therefore Meta analysis within the biosimilar context appear to be normal.	*between innovator and biosimilar (if there is a difference). Meta-analyses can help to more accurately define the effect or the originator compound (i.e. before starting the clinical comparability exercise of originator versus biosimilar).* • *According to the guidelines of biosimilar, the reported data of the established reference biologic was very important for the biosimilar developer to understand and evaluate the outcomes of comparison exercise of the candidate biosimilar.*
Do you think that for biosimilar comparability exercise the classical phase I, phase II and phase III approach is required? Please give your rational for or against that.	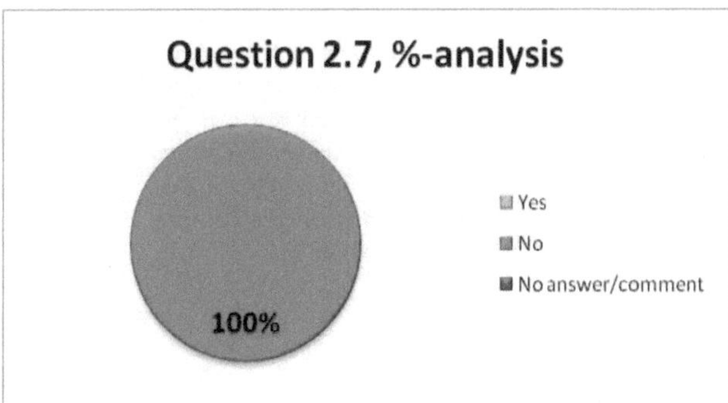

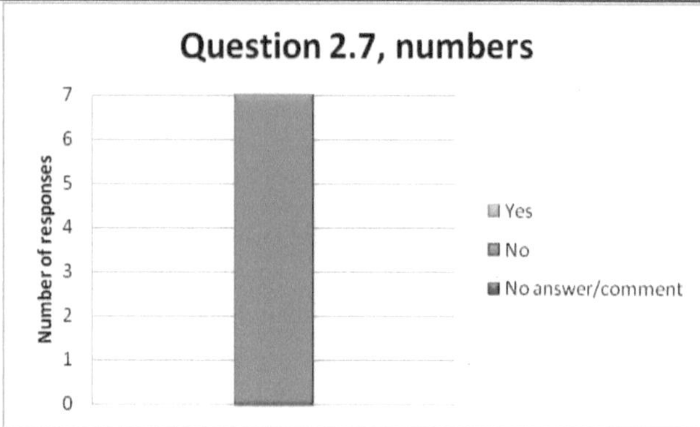

Additional comments from participants:

- *No as you now understand the disease and dosage so should just be comparing based on know efficacy endpoints*
- *Direct comparison should suffice as risk profile is known*
- *Absolutely not. Phase 1 is to show equivalent PK and PD if suitable PD markers. No phase ii is required and phase iii is designed to show equivalence not re-prove efficacy meaning different endpoints eg pd or surrogate endpoints can be accepted.*
- *See EMA guidelines*
- *The comparability exercise demands proof of PK equivalence and comparable efficacy/safety. Clinical trials should be designed to meet these objectives. Classical development (i.e., start to identify safe dose-range in humans, and do dose-finding trials before initiating confirmatory trials) is therefore not needed. Depending on the biosimilar product in question, clinical trial developments can be designed accordingly (case-by-case basis). EMA has even left open the possibility to do a comparative PK/PD exercise which can replace the need for comparative efficacy data.*
- *Based on the thoroughly comparability exercise, clinical data could be reduced for the development of biosimilar. For the clinical comparison exercise, in most of the cases, elaborate PK/PD studies will be enough before initiating the phase III.*
- *Studies in healthy volunteers are not always possible with existing standard of care, classical clinical trials approach may not be ethical.*

Would it make sense to test multiple lots of the reference biologic against multiple lots of the biosimilar during the clinical PK/PD studies?

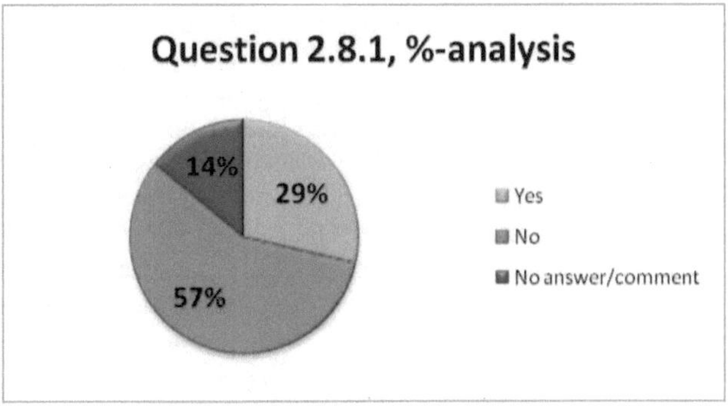

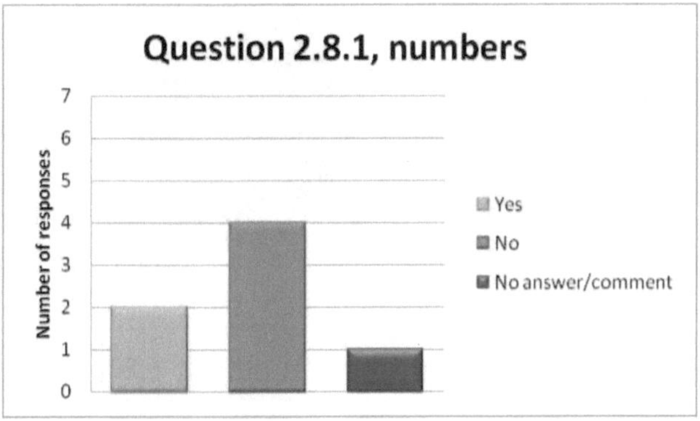

Additional comments from participants:

- *No this would add cost and complexity. Plan should be agreed with EMA up front.*
- *Generally, no - clinical trials are not sensitive enough to pick up differences between batches. Lot to lot variation should be controlled by sufficiently tight specification limits. There may be some circumstances where this could be a good idea eg where there is a narrow therapeutic window.*
- *Ideally yes, but not always meaning full*
- *If products are released within their specifications, they should not differ in efficacy/safety (and therefore PK characteristics) (otherwise the specifications are not adequate). Therefore, we must assume that all innovator product batches on the EU market have similar safety/efficacy, and PK properties.*
  With regard to the biosimilar: If the biosimilar is released using the specifications based on comparison to the innovator data, then the same applies to the biosimilar as well (no differences in safety/efficacy/PK/PD expected). The clinical

<table>
<tr><td></td><td>data (PK, PD, efficacy and safety) are confirmatory to the bioanalytical data in this regard.
• Since the safety and efficacy of the reference biologic had been proven by the clinical use. In the PK/PD and the needed clinical comparability exercise, only one lot of the reference biologic might be enough used as a comparator for multiple lots of the biosimilar. However, the parallel group design and multiple dose design may be necessary.</td></tr>
<tr><td>Please comment if you would combine PK, PD studies (step 1) with clinical efficacy studies (step 2)</td><td>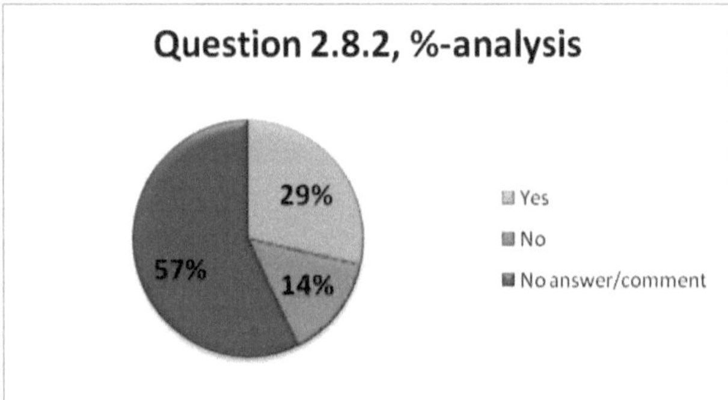

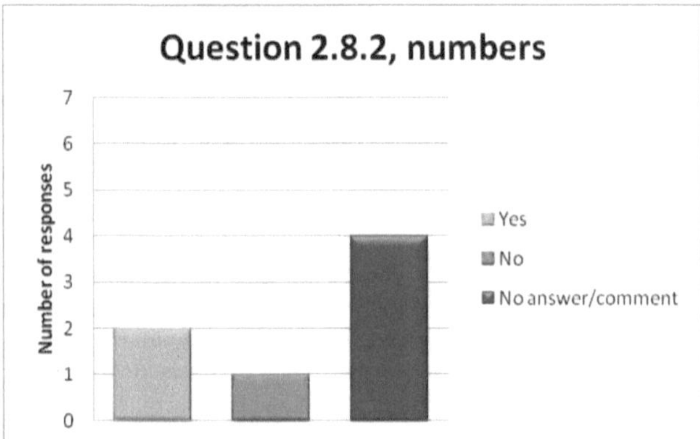

Additional comments from participants:
• I believe that was my idea. One needs to extract the max amount of data from patients. In some circumstances particularly oncology it makes sense to do this. There is the need for safguards and precautions and evaluation of data by an independent data monitoring board. Bear in mind there is a finite pool of patients and many competing trials.</td></tr>
</table>

	• *Beyond my expertise* • *No single answer can be given; the design of the clinical development plan varies on a case-by-case basis.* • *According to the step-by-step principle of development of biosimilar, results of one step should be evaluated before the beginning of the next step. Whether keeping the next step of development under the pathway of SBP or not depends on the judgement of the results of last step.*
Do you think that almost always the steps 2 (clinical efficacy) of the comparability exercise is required?	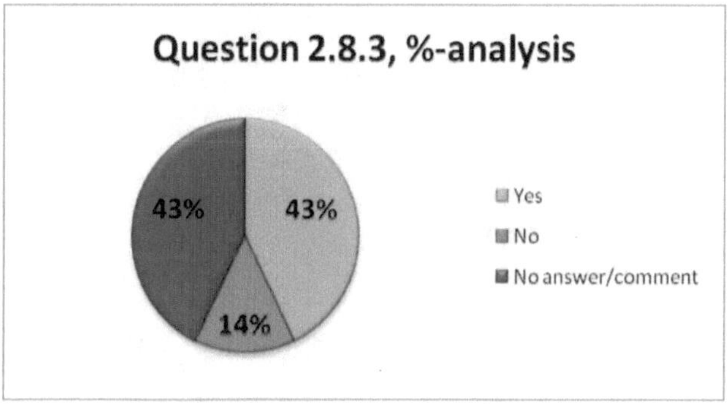 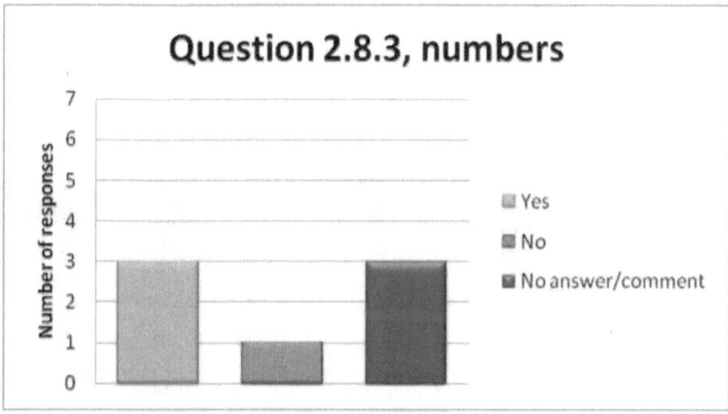 **Additional comments from participants:** • *No one can sometimes extrapolate efficacy from pd data as was done for zarzio. But still adequate safety and immunogencity data are required from representative patients.* • *Beyond my expertise* • *For the time being, regulators will have to get comfortable to approving biosimilars, and therefore they will usually require efficacy/safety data. However, there are definitely products with an adequate PD marker than may evade efficacy*

<table>
<tr>
<td></td>
<td>
endpoints.
<ul>
<li>In many cases, PD studies cannot be performed convincingly showing comparability in a clinically relevant manner. Therefore, similar clinical efficacy should be performed.</li>
</ul>
</td>
</tr>
<tr>
<td>Do you need to conduct this clinical efficacy exercise as a head to head comparison?</td>
<td>
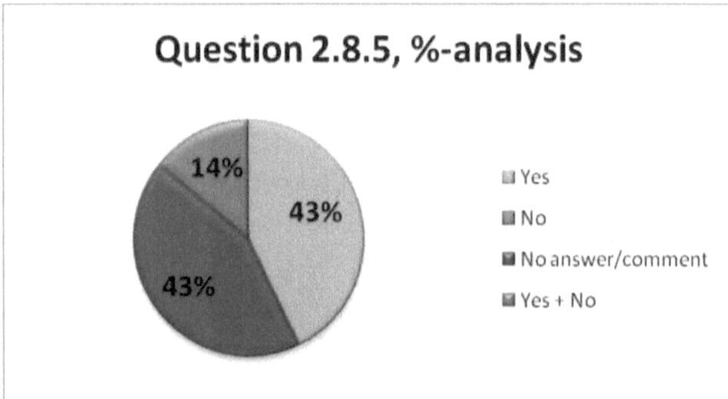

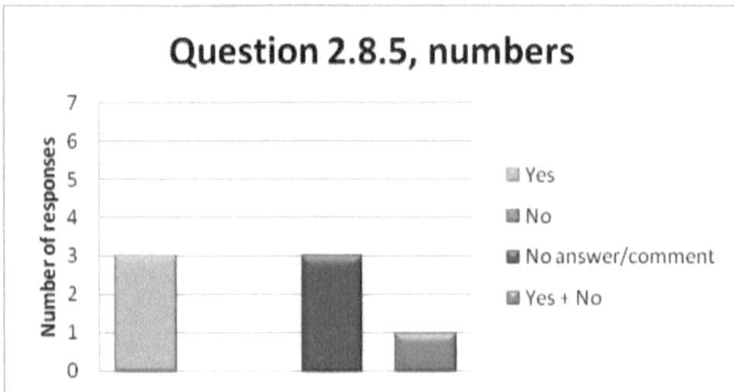

Additional comments from participants:
<ul>
<li>Usually yes but when efficacy is very clear/absolute egfor factor viii – then comparison against historical data can suffice</li>
<li>Beyond my expertise</li>
<li>As specified in the guidelines.</li>
<li>This is the basis for comparability exercise.</li>
</ul>
</td>
</tr>
</table>

<table>
<tr>
<td>Do you test multiple lots of the reference biologic against multiple lots of the biosimilar during the clinical efficacy exercise?</td>
<td>

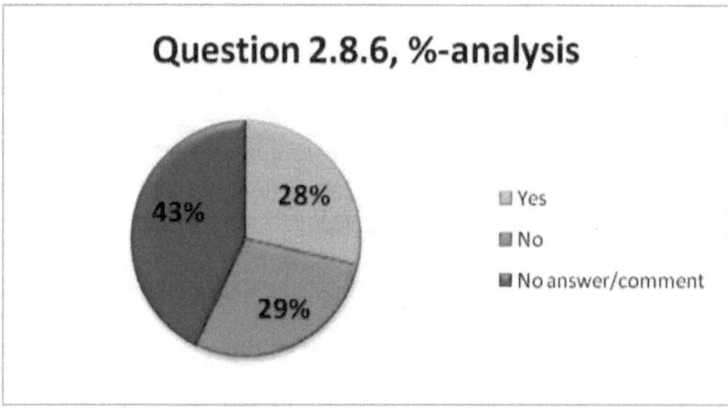

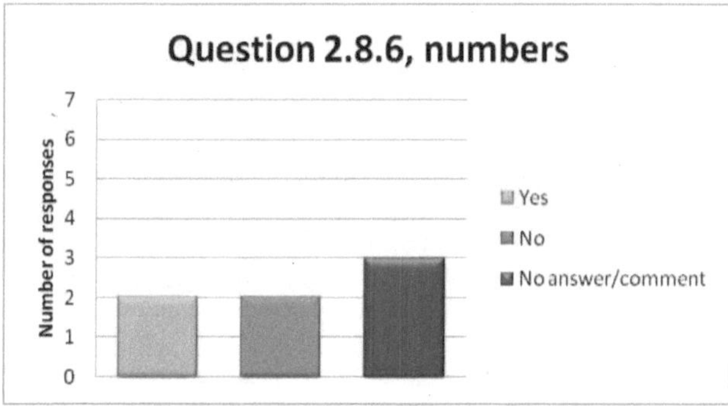

Additional comments from participants:

- *For reasons given for pk/pd but even more so no phase iii trial can discern between batches*
- *Inclusion of multiple lots in the efficacy trial will happen in large clinical trials where the number of patients to be treated, and the duration of treatment, exceeds the size of the batches (originator and biosimilar). However, the efficacy exercise does not demand that comparison of efficacy is made between different batches of originator/biosimilar. Also, the answer under 2.8.1 applies here as well.*
- *Only one lot of the reference biologic might be enough used as a comparator for multiple lots of the biosimilar. However, the parallel group design and multiple dose design may be necessary.*

</td>
</tr>
</table>

5.4 Discussion

The main finding with regards to the research question: "**what should be the scope of trials for biosimilars?",** of this study can be summarised in the following points:

1. **The trial size for biosimilars varies between 90 to ca. 500 subjects on average.**
2. **There have been no phase III trials finalized for mAbs. Current mAbs trials are on hold, due to differences in the views between developers and regulators on what is required.**
3. **The extent of the clinical trials depends on the extent of the similarity of the Q-profile.**
4. **It is disputed between experts, if multiple lots of the biosimilar require testing against multiple lots of the reference biologic during phase III trials.**
5. **Possible meta analysis within the biosimilars context is disputed amongst experts.**

5.4.1 The trial size for biosimilars varies between 90 to ca. 500 subjects on average

Based on the review of the literature, as outlined in section: 4.3.1 Results from the literature, the trial size varied for different products: more recently approved biosimilar products within the EU seem to have a trial size of ca. 500 subjects. It has been reported that the average trial size for increasingly complex biosimilars tends to be bigger than for relatively simple biosimilars, such as the ones first approved, for example, Omnitrop.

However, this answer, derived from the practical experiences with biosimilars, does not included mAbs, as those are not yet approved, some reports suggest that bioequivalence for a mAb (Infliximab) is expected to be achieved within a single dose trial. On the other hand, the development program for other mAb biosimilar appears to have been halted in phase III for TEVA and Samsung, refer to section 5.3.1 Results from the literature.

A possible explanation for this "mixed situation" is that the trial size varies due to the complexity of the biologic itself (publications suggest that the scope of clinical trial requirements for highly glycosylated proteins such as mAbs is more complex than initially anticipated by biosimilar developers) in combination with ethically justifiable trial settings.

Given that it is recommended in the new proposed guide: *"Guideline on similar biological medicinal products containing biotechnology-derived proteins as active substance: quality issues (revision 1) Draft,"* a series of successive steps, which build one on another should be used, starting with comparative PK and PD studies. PK studies, which should focus on detecting possible differences, comparative clinical trials will be necessary to demonstrate clinical comparability. There is general agreement about what clinical trials are required, refer to sections 5.3.1 Results from the literature and 5.3.3 Results for the survey.

5.4.2 The extent of the clinical trials depends on the extent of the similarity of the Q-profile

It is disputed between experts whether multiple lots of the biosimilar require testing against multiple lots of the reference biologic during phase III trials.

The answers to the research questions:

> **Do you test multiple lots of the reference biologic against multiple lots of the biosimilar during the clinical efficacy exercise?**

reveal that it is disputed whether multiple lots need to the tested as part of a clinical trials setting. Given that the impact of heterogeneity and variation of the quality profile cannot be estimated, clinical trials are required. However, if those trials are required for multiple lots is not clear.

This is also related to the question as to whether it is expected to detect differences in those trials, which is based on the trial design and the margins of tolerance, which set the boundaries for possible differences.

- *Equivalence trial clinically meaningful differences including superior efficacy between the biosimilar and the reference product are excluded if the 95 % confidence interval of the treatment difference is fully contained within the pre-specified two-sided (upper and lower) comparability margins.*

- *In the case of a non-inferiority study, only the lower margin is defined, i.e. the statistical analysis can only show that the product in not inferior to the reference product, but cannot exclude the possibility that the biosimilar product would be better than the reference product.*

The number of patients needed for an equivalence trial tends to be larger than that for the non-inferiority trial and the margins of tolerance were suggested to be 80 to 125% in *"Guideline on similar biological medicinal products containing monoclonal antibodies Draft"*. Margins of tolerance for the endpoints are controversial.

The third point to consider when talking about biosimilar trials is that they do not require a phase II, as agreed by all experts, refer to section: 5.3.3 Results for the survey, and question:

> *Do you think that for biosimilar comparability exercise the classical phase I, phase II and phase III approach is required?*

to which a clearly negative answer was given.

Only phase I and phase III trials were mentioned in EMA guidance, refer to section 5.3.1 Results from the literature. This makes it clear that phase II trials were not required. Further, there is the option to combine trials in such a manner as to combine various clinical setting efficiently. The questions:

- *Do you think that almost always the steps 2 (clinical efficacy) of the comparability exercise is required?*

- *Do you need to conduct this clinical efficacy exercise as a head to head comparison?*
- *Would it make sense to test multiple lots of the reference biologic against multiple lots of the biosimilar during the clinical PK/PD studies?*

were answered controversially, probably as it is recommended by EMA to conduct a step wise approach.

5.4.3 Possible meta analysis within the biosimilars context is disputed amongst experts

Meta analysis is disputed based on the output from questionnaires 1.1 and 1.2; however, within the biosimilar context this seems to occur, refer to section 5.1.2 Meta analysis for biosimilar clinical trials? (93).

Chapter 6.0: Why is extrapolation of indications for biosimilar controversial?

6.1 Introduction

Innovator medicinal products are approved and restricted by the competent authority so they are prescribed only for situations in which they have been shown, in dedicated clinical trials, to be safe and effective. Usually efficacy (associated with safety data) has to be shown for each indication separately in a clinical trial setting in a relevant patient population. This is also the reason why clinical trials for innovator medicinal products are extensive.

Extrapolation of indication is to be understood within the clinical trial context of biosimilars during the comparability exercise, to extrapolate safety and efficacy findings from one indication to another; however, with limited or no additional clinical trials.

The ability to extrapolate indicators was first introduced through the draft Guideline on similar biological medicinal products containing monoclonal antibodies Draft EMA/CHMP/BMWP/403543/2010.
As part of this study, triggered through the mentioning of extrapolation of indications in the above mentioned reference, this question was further analyzed.

The controversy related to extrapolation of indication is related to different clinical trial settings when conducting clinical trials for biosimilars (also refer to Chapter 5.0: What should be the scope of trials?). One major controversy is that there is not sufficient or no efficacy shown for the biosimilar for other indications with the exception of the one that is part of the comparability exercise. Given, the outlined dynamic of the profile and the issue of variation and heterogeneity outlined in Chapter 4.0: Main research question, there is the question of if the biosimilar, even if it has shown similarity for one indication, will be similar in other indications compared to the reference biologic.

For example, the mechanism of action of a mAb can be different for different indicators. Showing comparability for one mechanism of action is not sufficient for extrapolating efficacy for an indicator that exploits a different mechanism of action of the mAb.

6.1.1 Hypothesis: Extrapolation of indication can only be allowed if the mechanism of action is identical, if the mechanism of action is different some limited trials should be required

The mechanism of action might be different for different indications, as proposed by the EMA in their draft for mAbs, refer to Concept paper on the development of a guideline on similar biological medicinal products containing monoclonal antibodies. The adopted draft was Guideline on similar biological medicinal products containing monoclonal antibodies. Industry generally regarded this approach favourably.

The potential weakness is that the heterogeneity and variation of the biosimilar possibly has less impact on the efficacy of one indicator compared to the others, even if the mechanism of action is identical. This is a weakness that will be spotted and evaluated over time by the EMA. However, this case was not part of the scope of this study.

The overall approach of extending clinical trials is surely correct and it is not anticipated.

6.2 Experimental procedure (methods and materials) employed

6.2.1. Literature research

Publications were searched in MEDLINE (Pubmed) and Gabi-online. Key search terms were combinations of the words biosimilar and heterogeneity or variation or life-cycle. Searches were limited to full-text English-language articles. The websites from the European Medicines Agency (EMA) were also systematically consulted and selected biosimilar guidance documents were analyzed, refer to section: 3.3.1 Practical aspects of the questionnaires and critical considerations, for details.

6.2.2. Questions and comments submitted to EMA

Questions and comments were submitted to the EMA utilising the form and contact details provided by them, refer to Appendix: 1 Questions and comments submitted to the EMA and Appendix: 2 Questions and comments submitted to the EMA.

- The questions and comments for the Guideline on similar biological medicinal products containing monoclonal antibodies Draft, were submitted on the 30.05.2011, Appendix: 1 Questions and comments submitted to the EMA and Appendix: 2 Questions and comments submitted to the EMA and the EMA published answers to those on the 30.05.2012, available under Overview of comments received on 'Guideline on similar biological medicinal products containing monoclonal antibodies'.

Table: 27
Extracts from questions and comments submitted to the EMA. Refer to Appendix: 1 for more details.
The questions and comments for the *"Guideline on similar biological medicinal products containing monoclonal antibodies"* were submitted on the 30.05.11 and the EMA provided responses to those questions and comments, refer to *"Overview of comments received on 'Guideline on similar biological medicinal products containing monoclonal antibodies"* (69).

Question and comments submitted to EMA extracted from Appendix: 1	Notes and clarification in relation to research question
Extrapolation of indication for a Biosimilar However, as in any progressive document, there are a	The EMA wrote on this topic the following:

<table>
<tr>
<td>few areas that require further clarification. Although these areas are not specifically linked to mAbs and moreover to Biosimilars in a more general way, this guide uses some key terminology and concepts such as:

1. Extrapolation of Indications
2. Meta Analysis
3. Conventional Equivalence Margin

that require further explanation.

I feel it would be appropriate either to clarify those terms and concepts as part of this guide or to address them in a separate guide, applicable for all Biosimilars taking into account the heterogeneity and the nature of mAbs (and other Biosimilars).</td>
<td rowspan="2">"1.)
EMA/CHMP/BMWP/86289/2010 Guideline on immunogenicity assessment of monoclonal antibodies intended for in vivo clinical use. Draft

"MAbs exert their action by various mechanisms ranging from simple binding to antigen, which alone mediates the clinical effect, to binding antigen and mediating one or more immunobiological mechanisms which combine to produce the overall clinical response. Therefore, although simple binding may seem to be the only mechanism operating to achieve clinical efficacy, other effects may also play a role in this."

And

"The mode of action of the mAb (e.g., lytic, apoptotic), and especially the nature of the target molecule (e.g., immunosilencing, immunostimulating), needs to be adequately characterized and comprehensively investigated."</td>
</tr>
<tr>
<td>"Extrapolation of Indications
425 Extrapolation of clinical efficacy and safety data to other indications of the reference mAb, not
426 specifically studied during the clinical development of the biosimilar mAb, is possible based on the
427 overall evidence of biosimilarity provided from the comparability exercise and with adequate
428 justification".

Comment:
Referring to lines 424 to 428, what would the justification be? As the mechanism of action of the Reference Medicinal Product could be different from indication to indication, is it justified to allow "Extrapolation of Indications" without showing the efficacy for this particular indication for the Biosimilar?

For example, for some indication an mAb might bind to a cell surface, whereas in other indications the primary action is to bind to soluble antigen. The subsequent clearance and inactivation mechanism might be very different.

Differences in the glycosylation pattern, for example, between the Reference Medicinal Product and Biosimilar could be for one mechanism of action negligible, whereas for another mechanism of action very</td>
</tr>
</table>

important. The mechanism of action can be different from indication to indication. As the mechanism of action can be influenced by the glycosylation pattern, which can be influenced by the "Quality Profile" (refer to "overlaying issue of product heterogeneity and variation [(1)]"); an important consequence of this train of thoughts is that new clinical efficacy data would be required for a new indication in order to ascertain that efficacy of the Biosimilar is comparable to the efficacy of the Reference Medicinal Product for this new indication (assuming the case of a different mechanism of action).	

Refer to Appendix: 1 Questions and comments submitted to the EMA and Appendix: 2 Questions and comments submitted to the EMA for more details.

6.2.3 Survey research

A series of questions related to the research question was made public as a survey in selected forums as part of questionnaire 1.1 and was submitted as part of questionnaire 1.2 to selected stakeholders that included regulatory agencies, the innovator industry, the generics industry and consultants as well as other independent experts.

The following question relevant to the research question investigated was within questionnaire 1.1, refer to Appendix: 3 Questionnaire 1.1 for details.

Table: 28
Extracts from *"Questionnaire 1.1"*, refer to Appendix: 3, for more details.
The questions were published online in a dedicated forum, refer to section: 3.3.1 Practical aspects of the questionnaires and critical considerations.

Question from survey monkey questionnaire 1.1	**Notes and relation to research question**
Even if the mechanism of action is identical, could the efficacy be different as the isoform ratio may be different between biosimilar and reference biologic?	Yes / No type question

The following questions relevant to the research question investigated were within questionnaire 1.2, refer to Appendix: 4 Questionnaire 1.2 for details.

Table: 29 Extracts from *"Questionnaire 1.2"*, refer to Appendix: 4, for more details. The questions were sent to dedicated experts, refer to section: 3.3.1 Practical aspects of the questionnaires and critical considerations.

Questions from questionnaire 1.2	**Notes and relation to research question**
Extrapolation of indications. EMA suggests extrapolating indications if the mechanism of action is identical. Do you know of any biosimilar cases where this happened? Please cite, if possible.	Open question
Even if the mechanism of action is identical, could the efficacy be different as the isoform ratio may be different between biosimilar and reference biologic?	Yes / No type question

6.3 Results from why is extrapolation of indications for biosimilar controversial?

6.3.1 Results from the literature

The regulatory literature was screened as part of section: 2.3 Screening the above presented literature related to current biosimilar regulation with regard to the research questions.

The following regulatory literature topics impact on the research questions as summary in table form. For the detailed text passage, please refer to section: 2.3 Screening the above presented literature related to current biosimilar regulation with regard to the research questions and the following that contains comments from the author and further clarification.

Table: 30 Extracts from section: 2.3 series. Highlighting topics and findings from the analysis of regulatory documentation with regard to the research question. For the detailed text passage, please refer to section: 2.3 series and the following, which contains comments and further clarification.

Summary of regulatory literature review, impacting research question series 3 / Notes related to research question	**Refer to section**
Extrapolation of indication: *"It is important to consider the risk of immunogenicity in different therapeutic indications separately".*	2.3.7 Guideline on similar biological medicinal products containing biotechnology-derived proteins as active substance:

Conclusion/Comment: • This could relate to the issue that different patient groups with different underlying health issues could be treated with one biologic, but under different indications. Thus, a separate assessment might be required that the biosimilar is safe for each group of patients, depending on their respective conditions that are typical for their disease type. **Extrapolation of indication depends on also on safety aspects of subpopulation:** *"Possible safety issues in different subpopulations should also be addressed".* **Conclusion/Comment:** • Extrapolation of indication depends also on the safety aspects of the subpopulation.	Non-clinical and clinical issues EMEA/CHMP/BMWP/42832/2005 (55)
Extrapolation: For extrapolation of indication the following was mentioned: *"When efficacy and safety has been shown in one indication, extrapolation to the other indications is already now possible under certain circumstances.EMA/CHMP/BMWP/572828/2011 3/5. The BMWP intends to discuss these criteria and potentially revise or amend them with the experience gained so far".* *"The extrapolation to other indications which are not investigated during the clinical development program of the biosimilar product will be discussed".* **Conclusion/Comment:** • EMA is not fully sure about extrapolation criteria and awaits more (if any have happened so far) cases of extrapolation. Thus showing the attitude of the EMA, with regards to extrapolation, is adaptive.	2.3.8 Concept paper on the revision of the guideline on similar biological medicinal products containing biotechnology derived proteins as active substance: Non-clinical and clinical issues EMA/CHMP/BMWP/572828/2011 (56)

- An additional review of non-EMA papers and articles with regard to the research question was carried out in section: 3.1.2 Literature review.

Heterogeneity and variation and its impact on extrapolation

Dr. Kresse highlighted again in 2009 the role of glycosylation for mAbs. Even minor differences in glycosylation patterns can have significant pharmacological impacts. The pattern of glycosylation is influenced by the biologic itself (heterogeneity) and the process that is used to manufacture it (variation). With regard to extrapolation, "*the impact of glycosylation differences on clinical properties, which may be different for mAbs using different modes of action, should be proven or disproven*". (70) The glycosylation pattern (fucosylated[66] variants, for example) has an impact on the comparability of Fc-gamma Receptor binding and ADCC[67]. (9)

Conclusion/Comment

In line with the research question of Chapter 4.0: Main research question, there is an impact of heterogeneity and variation on the ability to extrapolate indications without additional clinical trial data. The better/closer the quality profiles of the biosimilars match those of the biologic, including purities and impurities, the better the rational for extrapolating from one indication to another is.

All functional parts need to be analysed

mAbs are complex biologics with various functional sites such as one Fc-part[68] and two Fab-parts[69], allowing for multiple interactions/interconnecting with target molecules in a complex cascade of events. Therefore, extrapolation of clinical safety and efficacy from one indication to another must be considered only when the justification for that includes an analysis of the effects of Fc and Fab sub-functions on the mechanism of action of the monoclonal antibody. (26)

Conclusion/Comment

The full complexity of the mAb needs to be understood, which exceeds the requirement of understanding the functional parts of the mAb only. Moreover, it is required to analyse thoroughly all functional aspects, even those that are not directly involved in the mechanism of action.

Extrapolation is not sufficient

---

[66] Fucosylation:
Fructose, as is the case for other reducing sugars, undergoes the Maillard reaction with proteins and amino acids.
http://www.ncbi.nlm.nih.gov/pubmed/8213610
The process of adding fructose sugar units to a molecule. It is a type of Glycosylation.
http://en.wikipedia.org/wiki/Fucosylation

[67] ADCC:
Antibody-Dependent Cellular Cytotoxicity. http://acronyms.thefreedictionary.com/ADCC

[68] Fc part:
The fragment crystallisable region (Fc region) is the tail region of an antibody that interacts with cell surface receptors called Fc receptors and some proteins of the complement system.
https://en.wikipedia.org/wiki/Fragment_crystallizable_region
http://www.bioatla.com/wp-content/uploads/Appendix_antibodystructure.pdf

[69] Fab-part:
The fragment antigen-binding (Fab fragment) is a region on an antibody that binds to antigens.
http://en.wikipedia.org/wiki/Fragment_antigen-binding
http://www.bioatla.com/wp-content/uploads/Appendix_antibodystructure.pdf

Some experts see extrapolation within the biosimilars context as genuinely critical. In general, biological activity per se can be assessed in non clinical settings such as cell-based assays and animal models (however the predictability of those usually requires confirmation). Even though this data eases the burden of clinical trials, it cannot be used to extrapolate biological activity in humans. *"Therefore controlled clinical trials are the most reliable method of showing similar biological activity between a biosimilar and originator product".* (19)

Conclusion/Comment

In principle, this train of thought is valid for any extrapolation, even if the biological activity has been shown to be similar in non-clinical and clinical settings. The question is to do with how the efficacy was established in the first place by the innovator company. The efficacy margins against a placebo arm may very well show in one indication a strong effect, whereas in the other one, the effect is of a much smaller magnitude. Therefore, the potential impact on similarity in one indication cannot be extrapolated for the other, as the margins of the effects against a placebo arm remain unknown. However, these should be taken into account when extrapolating from one indication to another, refer to note below. Thus, the clinical effects of the two biologics are compared in one indication. This head to head comparison might have different outcomes depending on the amplitude of effects in different indications.

Note: The biosimilar approach does not require a placebo arm to show efficacy per se as outlined in Chapter 5.0: What should be the scope of trials?; however, it is required, using sensitive endpoints, to show that the efficacy of the biosimilar is comparable to reference biologic.

Extrapolation is acceptable

Some experts see extrapolation within the biosimilars context as genuinely important. Dr. Ahmed stated that:

> *"Extrapolation of efficacy for a mAb from one indication to another without additional clinical trials is acceptable, based on comparability in physicochemical, non-clinical, and in vitro functional characteristics; bioavailability and clinical PK/PD of the one indication.*
>
> *"Once comparability has been demonstrated in one indication, there is no scientific reason to expect that the response of the host to the biosimilar product should differ from that of the reference mAb in other indications."* (70)

It is however acknowledged that safety can differ from one indication to another due to concurrent treatments with other medicinal products (cancer setting); however, even in those cases extrapolation of safety is also acceptable, assuming the indication selected for the comparability exercise is judged to have a high sensitivity toward detection of differences. Also, the patient population is acknowledged to have an impact on safety. It is suggested that the most sensitive population is selected for clinical trials (i.e. with potentially high response rates, if ethically justifiable) and/or those which are homogeneous and where biomarkers can be used during the comparability exercise. Further, it is suggested that additional immunogenicity concerns could be addressed in the post-approval phase (i.e. phase IV, refer to Chapter 5.0: What should be the scope of trials?) with regard to their impact on the patient populations. (70)

Conclusion/Comment

Extrapolation of efficacy is regarded as general possibility based on a successful non clinical comparability exercise, for the cases where the mechanisms of action are identical. Safety can be extrapolated based on the most suitable population in the detection of potential differences in the safety profile; even this will be difficult to justify scientifically as risks remain and the EMA is potentially not willing to wait for the additional safety data, during a phase IV period and prior to granting a M.A.

Extrapolation depending on the understanding of the mode of action

As mAbs are multifunctional biologics, their mode of action may depend on, for example, the type of target.

- If the mAbs antigen[70] is a soluble antigen, then it is more likely that there is only one mechanism of action and extrapolation would thus appear feasible.
- If the mAbs antigen is cell-bound (for example, a receptor molecule), then several mechanisms of action can be involved. For example, blocking the target receptor and triggering signalling events that initiate the immune cells to attack pathogenic cells to which the receptor is bound (73). Characterization of the quality profiles and preclinical data may facilitate extrapolation.

Conclusion/Comment

Extrapolation between indications is possible; however, the understanding of the mode of action is critical.

6.3.2 Results from questions and comments submitted to EMA

Refer to Appendix: 1, Appendix: 2 and Appendix: 5 for details.

1) The questions and comments for the *"Guideline on similar biological medicinal products containing monoclonal antibodies"* were submitted on the 30.05.11 and the EMA provided responses to those questions and comments, refer to *"Overview of comments received on 'Guideline on similar biological medicinal products containing monoclonal antibodies"* (69), also represented in Appendix: 5.

Result:

- No relevant answers were given by the EMA to the submitted questions and comments.

[70] Antigen:
A toxin or other foreign substance which induces an immune response in the body, especially the production of antibodies. Oxford Online Dictionary.

2) Questions and comments for the *Guideline on similar biological medicinal products containing biotechnology-derived proteins as active substance: Quality issues (revision 1)* were submitted on the 28.11.2012. (A response from the EMA is still outstanding in April 2013.)

6.3.3 Results from the survey

The following findings were summarized from questionnaire 1.1. For more details, refer to Appendix: 3 Questionnaire 1.1. In cases where a potential controversy was observed all comments from the participants were included in this study. Whereas, in the cases where an overwhelming majority was reached, no further comments were included.

Table: 31
Results from *"Questionnaire 1.1"*, refer to Appendix: 3, for more details.
The questions were published online in a dedicated forum, refer to section: 3.3.1 Practical aspects of the questionnaires and critical considerations.

Questions from survey monkey questionnaire 1.1	Results from research question
Even if the mechanism of action is identical, could the efficacy be different as the isoform ratio may be different between biosimilar and reference biologic?	**Comments from participants:** • *http://www.surveymonkey.com/i/t.gif* • *General comment/suggestion. This is a very limited survey from which little information can be gathered. Also be aware that some questions and proposed answers are often ambiguous or incomplete. It also appears to be a mix of questions assessing attitude on the one hand and others assessing scientific knowledge on the other hand.* • *If the efficacy is different, then the product is not - by definition - a biosimilar* • *It is very possible. Different isoforms can have different potencies. Company needs to understand this.* • *The efficacy could be different scientifically and hence the need for clinical studies. If they are different, then I do not believe it should be approved as a biosimilar.*

The following findings were summarized from questionnaire 1.2. For more details, refer to Appendix: 4 Questionaire 1.2. In cases where a potential controversy was observed all comments from the participants were included in this study. Whereas, in the cases where an overwhelming majority was reached, no further comments were included.

Table: 32
Results from *"Questionnaire 1.2"*, refer to Appendix: 4, for more details.
The questions were sent to dedicated experts, refer to section: **3.3.1 Practical aspects of the questionnaires and critical considerations**.

Questions from questionnaire 1.2	Results to research question
Extrapolation of indications. EMA suggests extrapolating indications if the mechanism of action is identical. Do you know of any biosimilar cases where this happened? Please cite, if possible.	**Comments from participants:** • *Don't know?* • *No comment given.* • *Yes somatropin (Valtropin) for one. Filgrastim (zarzio) for another* • *Yes, see list of already approved biosimilar* • *Only few biosimilar compounds have been approved, I am not sure if any of these biosimilars applied for a waiver of indications.* • *No comment given.* • *Filgrastim*
Even if the mechanism of action is identical, could the efficacy be different as the isoform ratio may be different between biosimilar and reference biologic?	**Comments from participants:** • *Don't know?* • *No comment given.* • *If the moa is the same why should the efficacy differ between indications? However moa is not always fully understood – that is where the debate lies right now.* *Actually one needs to check that differences in isoform do not impact critical biological effects through binding and in vitro studies. Those are far more sensitive ways of detecting a difference. If a difference in activity is seen then more extensive clinical testing would be required.* • *In theory yes, however this is usually appropriately analyzed when relevant biological assays are available* • *I am not aware of efficacy changes due to isoform ratio-changes. I suppose that such differences and*

	their impact on PK, efficacy, and safety, would have to be assessed on a case-by-case basis. • *Because some isoforms of CQA such as the sialylation, might affect the clinical efficacy of the biologic by change the in-vivo half-life.* • *Seems plausible... Therefore, any differences detected at the quality level, need to be taken into account and may lead to more non-clinical and/or clinical studies.*

6.4 Discussion

1. **Extrapolation of indications is still a vague topic. It has not occurred for mAbs yet.**

As outlined in the introduction, section: 6.1 Introduction, extrapolation of efficacy is estimated to be difficult. The possibility of extrapolating safety, including immunogenicity data also requires careful consideration, and may have to involve more specific studies.

For example, mAbs have diverse functional activities and may be used in diverse indications; as in the case where the reference biologic (mAb) is approved as an immune-modulator and as an anticancer antibody. The antibody depended cellular cytotoxicity (ADCC) is a known mechanism of action for Abs, this mechanism of action can be required for one indication, however, signalling inhibition also results in the induction of apoptosis that can be a mechanism of action in an Ab.

Thus, different indications require different active properties of the reference biologic (including different involved sub compartments of an Ab, for example) and receptors or potential antigens of those Abs (or combinations of these).

- This is further complication by factors such as different sites of activity within the human body
  (i.e. different pharmacologic milieu) being involved. Also, the route of administration into those different milieus and the dosage might be different from indication to indication.

- Further, the patient population from one indication to another may be (very) different, spanning from relatively healthy patients with only immune-modulator problem to patients with a heavily degraded immune system due to chemotherapies, for example.

Thus, for example, a biosimilar mAbs, with similar effects against one disease compared to the reference biologic may have different effects in a second indication, if the second indication involves a different mechanism of action.

Further, even in the case where the mechanism of action is identical or similar between indications, the question remains, what is the understanding of this mechanism of action? The biosimilar company is only required to show comparability, assuming the mechanism of action was sufficiently explored and discovered by the innovator company. However, this remains and is an assumption.

Therefore, scientifically justifying extrapolation between two (or more) indications is controversial. Additional extensive quality and non-clinical data, including potency assay(s) and in vitro assays that cover all functional aspects of the biologic (for mAbs there are various functional aspects to consider), and upon the condition that the mode of action is identical (or highly similar; as the mode of action itself is often not fully understood for the reference biologic), could enforce the biosimilar claim for another indication. Further, immunogenicity cannot be predicted, so it must be measured directly in the most sensitive patient population. This could well be a different population compared to those used for the innovator biologic and to the one used to show clinical efficacy.

2. **Not only is the mechanism of action required to be identical, the patient population should be similar as well.**

It is required to assess if different patient groups with different underlying health issues could be treated with one biologic, but under different indications. Thus, a separate assessment might be required that shows the biosimilar is safe for each group of patients, depending on their respective conditions that are typical for their disease type. Section: 6.3.1 Results from the literature indicates and confirms that extrapolation of indication also depends on the safety aspects of the subpopulation.

3. **Experts mostly agree that the isoform and glycans structure are very important in the allow-ability for extrapolation of indications, without clinical trials.**

The full complexity of the biosimilar, including functional sub-parts, as it is the case of mAb, for example, needs to be understood, which exceeds the requirement of understand the functional part that is relevant of the mechanism of action.

Therefore, as indicated in Section: 6.3.1 Results from the literature, the impact of glycosylation differences on clinical properties, which may be different for mAbs using different modes of action, should be proven or disproven. Of course, this requires as outlined in Section: 6.3.1 Results from the literature a good understanding of the mode of action as a prerequisite.

4. **Outlook**

The situation with regard to extrapolation of indications seems to be still evolving as:

"*BMWP intends to discuss these criteria and potentially revise or amend* them with the experience gained so far. *The extrapolation to other indications which are not investigated during the clinical development program of the biosimilar product will be discussed".* (98)

Thus, there is a need to wait for more guidance and practical examples in this area.

Chapter 7.0 Integrated discussion

7.1 Part I: General Comments

In the introduction part of this study a series of unsolved questions with regard to the development of biosimilars were brought forward. Those questions were directly addressed in the following chapters. Chapter 7 will present the integration between the research questions and encompass the issues as well as represent and discuss them in a comprehensive way.

The main research question of this study, as per chapter 1, was outlined as being about the similarity of the biosimilar to the reference biologic in conjunction with the whole lifecycle of the two biologics. This resulted in directly related questions about the development strategy and possible solution encountered during the development: refer to research questions 2 and 3.

The underlying assumptions of the study about heterogeneity and variation of biologics were confirmed for biosimilars too, which results from a fact and mainstream-opinion based foundation.

1) Purity. Starting with ***purity***[71] of biologics in general, what does the term purity mean within the biologic context? If a protein is found to be highly pure, this does not mean in any way that it is so. Biologics are and remain complex mixtures with purity and impurity profiles, which are based on the resolution of the analytical techniques and are expected to show differences. The emphasis is that there are no identical protein molecules derived from a process, as already described in the biosimilar pathway; it is subject to flexible interpretation on what is regarded as "highly pure" or "pure enough". Further improvements in analytical techniques may show that innovator biologics, which were regarded years ago as being highly pure, now being more heterogeneous.

Experts agree that within one lot of a complex biologic there will always be variations of mAb molecules (due to product or process related heterogeneity), which are slightly different to each other, and that the ratio of those slightly different bio-molecules is unique for an individual lot. As a consequence, between multiple lots of a complex biologic (such as mAb), there will always be lot to lot variation of the Q-profile (due to product or process related variation and/or process changes), refer also to section: 4.3.3 Results from the survey research method.

The term "purity" is taken on form small molecule pharmaceuticals. For biologics, it is a plural term and should mean the summary of isoforms of the active substance, i.e. within the protein context a protein with the same amino acid structure but with different secondary modification such as the glycosylation.

---

[71] Purity:
Freedom from adulteration or contamination. Oxford Online Dictionary.

However, from a regulatory perspective, assuming that the reference biologic is not a "highly purified protein", what is the justification for removing it from the biosimilar pathway in such cases? In theory, the biosimilar pathway should be open to all biologics. That should also include very complex biologics such as vaccines, for example, which will be a considerable challenge for the regulator (EMA). (However, in practice there is no demand from the industry and this is not a likely case.)

In summary the mentioning of highly purified proteins takes likely practical consideration into account, associated to mimic quality profiles (as one main postulation forwarded and adopted by EMA, refer to chapter 4). This might be a challenge for the biosimilar industry and their technical abilities and know how, but cannot be regarded as a justified regulatory constrained, which is allowed to limit the biosimilar pathway to "highly purified proteins" only.

2) Consistency[72] is also a neglected word within the biologic context. However with regard to the biosimilar situation, this term requires some clarification. Firstly, 'consistency towards what?', is important within the biosimilar context. That is the question to pose. Batch to batch consistency is unclear, as each individual product and its unique production process will give rise to a range of complex bio-molecules that are difficult to fully characterize and to reproduce in their exact ratios and identities from batch to batch.

- If this means that batches/lots shall have a similar degree of heterogeneity and variation when compared to each other, this would accommodate the dynamic of the quality profiles. Consistency can be regarded as having a constant ratio, identity and quantity of purities and impurities between batches/lots. This is a more practical and advisable approach, as a development aim the consistency of the biosimilar is bound to the consistency of the reference biologic, implying the ratio of isoforms is consistent form batch to batch with consistent respective quantities, and bioavailability for each isoform within a batch.
- If consistency is considered as a standalone, one can understand the word consistency as the attempt to focus biosimilar development efforts towards one particular isoform for a biologic, for example. Thus, it is confounded with the term purity, which is controversial by itself, as discussed above.

The reference biologic company likely has to submit clinical data to justify the allowable ranges for what is considered as being consistent enough, and the clinical relevance of this consistency was tested; however, within the biosimilar context this is and will likely not be the case. Thus, from a regulatory perspective the required consistency of the biosimilar is found based on the consistency of the reference biologic.

A biosimilar with a lesser degree of heterogeneity and variation, i.e. better consistency, can be interpreted in the context of "highly purified proteins" as a more pure product with a less

[72] Consistency:
The quality of achieving a level of performance which does not vary greatly in quality over time. Oxford Online Dictionary.

dynamic quality profile. This would result in it being a bio-better, but not a biosimilar anymore.

Thus, this differentiation is of fundamental importance with regard to the envisaged approval pathway. Assuming the biosimilar pathway is envisaged, then consistency of the biosimilar with the reference to the innovator biologic should be used. Only the reference biologic dictates the boundaries for the biosimilar developer, the better those are characterized the wider the options are for the biosimilar developer.

Therefore, the biosimilar developer has to mimic the quality profile of the reference biologic, including all its purities and impurities (as postulated in chapter 4). It would be wrong just to mimic the major isoform, as different isoforms in different indications of the same biosimilar might have different clinical properties. A very recent publication followed the same train of thought:

> *"Structural differences between a biosimilar and its reference biological are only acceptable within the heterogeneity pattern of the molecule, and any differences found have to be explained and justified with regard to the potential impact on the clinical performance of the biosimilar".* (99)

For the biosimilar developer it is safer to aim for batch to batch consistency, which is not interpreted as the inverse of a dynamic of the quality profile. Batch to batch consistency should be viewed by the biosimilar developer as consistent purities and impurities, where for the purities it is of particular importance to have consistent qualities and types of isoforms.

3) Heterogeneity and variation

The above points all fall within the scope of research question series 1. They are all related to the fundamental issue that quality profiles change over time and its implication. The major implication for the biosimilar developer remains not addressed by the EMA.

The implications of this situation are that the dynamic development process of the biosimilar needs to be synchronised with the dynamic evolution of the reference biologic. This is a potential major problem within the biosimilar concept.

A biosimilar developer may be able to understand the heterogeneity and variation of the reference biologic very well; however, the risk always remains that those established ranges for the reference biologic shift at an unfortunate time point in the biosimilar development program. Potentially the most unfortunate time point is when clinical trials were being conducted. The EMA has made it clear that once the M.A. is granted for the biosimilar it is expected that both products may vary independently, however, prior to the M.A. being issued for the biosimilar the situation is not clear.

There are a couple of possible scenarios in which the dynamic of the quality profiles will cause problems during the biosimilar development (examples were submitted to EMA as part of Appendix: 4) and the current state of the biosimilar regulatory approach does not precisely describe what to do.

One aspect of the above is related to the sameness of the reference biologic. As elucidated above the heterogeneity and variation result in an ever changing quality profile of the reference biologic. Thus, it needs to be taken into consideration at non-clinical and clinical development time points of the biosimilar that the reference biologic also changes its quality profile, with dynamic extent.

This "unclearness" and the resulting flexibility on regulatory guidance on the topic can be viewed as a risk by the biosimilar industry and therefore as a discouraging obstacle for biosimilar development. This is a major constraint.

Projected planning should be possible for the biosimilar industry, related to development timeframes and associated costs. As the exact requirements on "similarity" imposed by EMA remain, depending on each of the scenarios, "flexible" and taking the dynamic quality profile issue and its implication for the development into account, the costs and success of a biosimilar development program are difficult to estimate.

4) Trials of biosimilars

Biosimilar Immunogenicity is an ongoing concern for all biologics and is more conclusively addressed during the clinical phases of the reference biologic approval, as the trial size is much bigger compared to biosimilar trials. It is more likely that during this clinical evaluation phase, multiple lots of the reference biologic have been tested, which is a means of testing the clinical impact of heterogeneity and variation, whereas the clinical trials for biosimilars seem to be limited to a head to head comparison, likely carried out with one lot of the reference biologic tested against one single lot of the biosimilar.

The other reason why the extensive safety data, which exist for the reference biologic, is of limited use for the biosimilar developer is because it is just partially representative of the current safety of the reference biologic. Thus, an extrapolation of old safety data from the reference biologic towards the biosimilar, once biosimilarity was obtained (at one given moment in time), does not mean that all old safety data available for the reference biologic can be utilised to justify safety for the biosimilar. On the contrary the biosimilar, due to its unique and thus different production process must show independent safety for patients by clinical trials, which cannot be abbreviated with historical safety data from the reference biologic itself. Otherwise patient safety would be put at unnecessary risk, which is not the aim of the biosimilar developer nor would such a dossier be approved by the EMA.

If biosimilarity is demonstrated during the comparability exercise, it cannot automatically be concluded that all the safety data obtained for the reference biologic is of relevance for the biosimilar itself. The reason for this is that both have their independent lifecycles and biosimilarity is just demonstrated at one given point. Thus, the EMA requires a comprehensive pharmacovigilance and risk reduction plan for the phase IV trials of biosimilars, i.e. post M.A. phase.

Biosimilar efficacy

The intention of the biosimilar development is to show similarity with the reference biological, not to independently demonstrate patient benefit. Therefore, the biosimilar is required to show therapeutic equivalence (within margins of tolerance that require justification) compared to the reference biologic. **This infers that the biosimilar does not have better or worse efficacy than the reference biological, thus allowing the use of the same dosage for the same indication.** This means that the trial scope should be based on an equivalence trial with lower and upper margins, contrary to a non inferiority trial.

That, at least, will allow for the same indication to switch from an originator biological to the respective biosimilar. The following will be identical:

- Indication
- Therapeutic regimen i.e. plan for treating the disease (e.g. intervals and dosage etc)
- Therapeutic dosage (concentration)
- Therapeutic form and related route of administration

5) Post-marketing

The need for strict post-marketing surveillance and potentially large, post-marketing studies to provide reassurance regarding the safety of biosimilars, particularly epoetins, has been highlighted. EMA requires that for a biosimilar, as for any new drug, a comprehensive risk management plan, including a plan for post-authorization safety surveillance, be submitted at the time of the marketing authorization application.

EMA actively and comprehensively follows the safety of biosimilars and tailors the post-marketing programme for a specific biosimilar according to identified and potential safety concerns for the biosimilar, the reference biological and/or the substance class.

7.2 Part II: Findings evaluation

1) The central question in this study was triggered by the issuing of EMA/CHMP/BWP/617111/2010, with emphasis on the life-cycle of the biosimilar; does the biosimilar need to achieve a quality profile that falls within the quality profile of the reference biologic? Under what circumstances could this approach be deviated from?

This is directly related to the development strategy a biosimilar developer should take given the close link on how to tackle this dynamic in a comprehensive development strategy for the biosimilar. The question of the development aim or strategy of a biosimilar is of vital interest for the biosimilar developer.

The following key postulations were made during this study and independently submitted to the EMA, refer to Appendix: 3 and Appendix: 4:

> *"The initial or cardinal biosimilar development aim is to mimic the quality profile of the reference biologic drug substance including its variation/heterogeneity (on a lot basis) where possible."*

"The better the correlation of the molecular characteristic (including process understanding) of the biosimilar drug substance with quality attributes is understood, the more the biosimilar developer can shift the focus (of the development strategy) from mimicking the Q-profile of the reference biologic, including its variability/heterogeneity, towards manufacturing a more consistent/pure biosimilar drug substance; and therefore, eliminating some of the heterogeneity of the reference biologic, as long as the safety and efficacy remain comparable within the margins of equivalence."

"The Biosimilar requires to such an extend similarity to the Reference Medicinal Product, as the Reference Medicinal Product is similar to itself, when compared on a batch to batch basis, as the optimum. Deviation of this approach is generally acceptable, but requires justification and possibly additional clinical evaluation in order to assess associated risks."

"Further, to guide the industry and avoid unnecessary costs and disappointments, it should be stated that:

For non clinical setting it is advisable that the "Quality Profile", (refer to [(1)]) of the Biosimilar is within the levels of variation observed for the "Quality Profile" of the Reference Medicinal Product."

"Thoroughly determination of the "Quality Profile" of the Reference Medicinal Product can be used to set the margins of tolerance (for variation) for the Biosimilar "Quality Profile". Deviation of this approach is generally acceptable, but requires justification and possibly additional clinical evaluation in order to assess associated risks".

They all summarise with different words that the reference biologic, including its heterogeneity and variation, should set the boundaries for the biosimilar and are essential for a coherent development strategy. Those postulations were partially adopted by the EMA, ultimately showing that they withstand examination. In the case where they were not adopted, this was related to duplication or not being postulated within the right category of guidance documents. However, no comments were made by the EMA stating they were found to be unreasonable. The majority of the participants of the questionnaire would adopt those postulations as well, setting a clear target for biosimilar developers.

Taken together, the data from this study suggest that it is required to mimic the purities and impurities of the reference biologic and to justify, if the biosimilar is less impure i.e. has a more pure quality profile with less heterogeneity and variation (for purities and impurities).

This has implications for all the 5 points cited in part I:

1) Purity
2) Consistency
3) Heterogeneity and variation

4) Trials of biosimilars
5) Post-marketing

2) The trial size is a major cost during the development of biologics. Thus, the interest of the biosimilar industry is naturally to keep trials limited. It can be regarded that the quality profile and clinical profiles are interlinked.

- For the safety aspect, immunogenicity is a critical point to consider as the biosimilar is derived from its own unique production process. Thus, extrapolation of safety data from the reference biologic is not feasible. The process related impurities, aggregation and unusual glycans must be addressed separately in clinical phase I trials. As a guide, it should be considered that a biosimilar should have a similar safety profile and as a consequence should not have a (significantly) better safety profile.

- For efficacy aspects, PK/PD it is a critical point to consider as the different glycan structures have in theory independent efficacies. The mechanism of action of a bio-molecule as well as its bioavailability is influenced by the glycans structure, thus all isoforms of the reference biologic in a similar ratio and quantity should be mimicked by the biosimilar developer.

Taken together, the quality profiles impact the clinical behaviour.

The need to synchronise the quality profiles:

When taking the context of the dynamic of the quality profiles into consideration, it is advisable, from the biosimilar development perspective, to synchronise the quality profiles before and during the comparability exercise, refer to Figure: 2.

Figure: 2

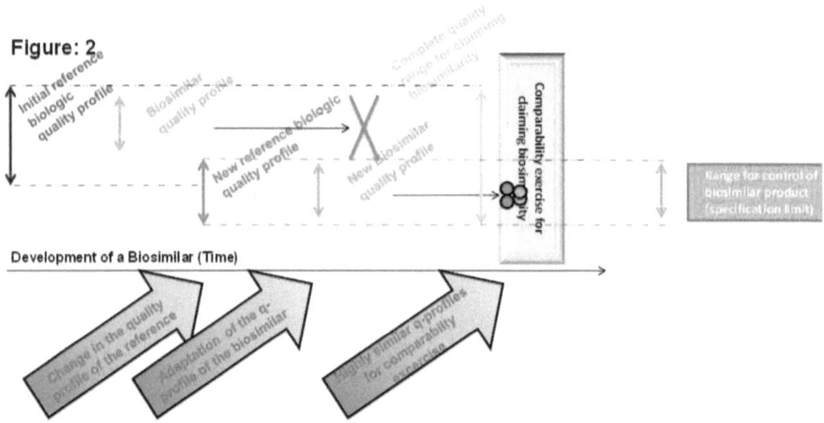

Figure: 2
The "Initial originator quality range" is the quality range of the reference biologic in dark blue.
The quality range of the quality profile of the biosimilar developer is in light blue.
Clearly, the quality profile of the biosimilar fits within the limits of the quality profile of the reference biologic.
Then in pink, the quality profile of the reference biologic has changed. The new quality profile is adopted by the biosimilar and the comparability exercise is carried out with highly similar quality profiles.
The allowable quality range (quality profile) is dictated by the quality profile of the biosimilar.

Whereas, the boundaries are set by the reference biologic, and the biosimilar should follow those closely, as required, until the M.A. is issued by EMA.

Of course It can be argued, (as some participants of questionnaire 1.2 did) that in the case where the quality profile of the reference biologic has shifted, clinical trials of both biologics with unequal quality profiles can still be conducted based on the assumption that only changes of the quality profile of the reference biologic, which have no (major) impact on safety and efficacy, will be approved by the EMA.

However, this perspective is misleading, as it does not take a series of issues into account:

- The accumulated historical clinical data for the innovator biologic spans multiple lots, which takes the impact on safety and efficacy of a greater portion of the quality profile into account. Thus, pre and post quality profile changes can be regarded as similar or be extrapolated based on this data. This is especially of relevance if the post change will result in an "improved" clinical profile, typically the aim of a change impacting safety.
- Extensive pharmacovigilance data is usually available for the innovator, which can be used in combination with historic changes (i.e. process knowledge gained by the innovator company) to the process, to make an argument for the change without the need to pass through extensive clinical trials.
- Especially for the case where extrapolation of indications is envisaged solely relying on the "EMA change control" and assuming that totally untested clinical behaviour is still similar while knowing that the changes in the quality profile of the reference biologic appears to be a risky strategy not only for the biosimilar developer but also for the EMA, as potential post MA issues will be due to a too lax approval process for biosimilars.

In summary, the innovator company is regarded by the regulator (be it justified or not) as knowing their product/process and the relationship between changes in the quality profile and their clinical implication better, due to the extended experience/knowledge gained.

The question of testing multiple lots

As a consequence of the rationales presented above that each lot can be regarded as having its own clinical profile, due to its unique quality profile. For the reference biologic it can be assumed that multiple lots were tested over the long years of standard clinical phases I, II and III trials. For biosimilars however, only very few lots will be tested against the reference biologic. One can estimate ca. 500 patients could be tested in a setting where one lot of the reference biologic was tested against one lot of the biosimilar during phase III trials. Preceded by a phase I trial for the biosimilar only with 100 to 150 patients. Thus a slim biosimilar development program can comprise only two lots.

Assuming that phase III trials show comparability of the biosimilar to the reference biologic, the question remains if that data can be extrapolated to other indications, as the quality profile has an impact on safety and efficacy, thus there is the question, what are the implications of heterogeneity and variation with regard to extrapolating indications?

An article published on the 29.04.2013 described that there are halts on clinical phases for mAb, in line with the reported halts in chapter 4. Thus, this indicates that the clinical trials strategy and setting is not clear to biosimilar developers, and the EMA probably recommended the change to the clinical trial strategy with the requirement for more extensive data. This speculation is supported by the article where Celltrion stopped a phase III trial of the biosimilar rituximab due to recommendations received from regulatory bodies (presumably the EMA, as the halted trials were conducted in the EU) as the phase I trial requires completion before proceeding to phase III. Thus safety concerns may have triggered that phase III halt (100).

3) Extrapolation of indications is still a vague topic and has not occurred for mAbs yet

As outlined in the introduction, extrapolation of efficacy is estimated to be difficult. The biosimilar industry has an interest in extrapolating indications; however, from a patient perspective it must be assured that PD/PK and immunogenicity are similar to the reference biologic for each indication.

Due to heterogeneity and variation of the biosimilar, it is possible that the purity and impurity profiles will be different compared to the reference biologic. Assuming that the impurity profile will have no impact during the comparability exercise; due to improved production technologies, the purity profile could be of bigger concern. Even if the mechanism of action is the same between indications, the different isoforms of the purity profile can still have an impact on the PD/PK behaviour, which for different indications can result in different clinical behaviour. For example, for different indications the milieu in which the mechanism of action is supposed to take place can be different. Thus the different isoforms of the purity profile with all their functional groups need to be analysed. If solely non-clinical data will be deemed as being sufficient remains questionable.

Further, for extrapolation of indication where non inferiority clinical trials were conducted, the evidence can be deemed as being not enough to allow for extrapolation of indication. As non inferiority in one indication with clinical data could be deemed by EMA as being acceptable; however, non-inferiority then extrapolated to other indications could mean that treatment effects are significantly different due to different ratios of isoforms, thus treatment effects could be (much) stronger in other indications.

In principle, the more complex the bio-molecules are the more risk is involved to mis-extrapolate (mis-estimate) safety and efficacy for another indication without any additional clinical trials, thus for mAbs it remains to be seen if extrapolation of indications will be carried out.

7.3 Part III: Discussion of solutions (and outlook)

7.3.1 Solution for development strategy of biosimilars

Taken together, sudden changes in the quality profile of the reference biologic can jeopardize the comparability exercise of the biosimilar in cases where those changes have an impact on the safety or efficacy. It is not foreseeable for a biosimilar developer if those differences will have an impact, as minor differences in the glycosylation pattern for mAbs, for example, can have a different PK and PD impact in cases of efficacy and immunogenicity for safety.

As mentioned before, the EMA does not give guidance on how to tackle those situations; however, during the course of this study two strategies were presented:

1) One approach is to demonstrate that those changes in the reference biologic's quality profile do not result in inferior clinical performance during the comparability exercise of the biosimilar:

> *"Changes in quality attributes may be acceptable if it is demonstrated that there are no negative impact on safety and efficacy. Drift of quality attributes, the quality profile could occur over time,*
> *Challenges for the biosimilar developer when defining the target variability of reference product attributes could occur.*
>
> *Dr. Ho mentioned for acceptable changes, that depending on the quality attribute and their potential impact on safety and efficacy, could support and justify differences observed at the time of submission."*

However, when employing the term non-inferiority for clinical trials, this is not in line with the biosimilar philosophy where equivalence trials are most appropriate. Problems will likely occur (depending on the quality profile) when attempting to extrapolate for indications.

2) The other approach would be to establish a design space[73]. If the biosimilar developer would establish a design space for the biosimilar, then it would be possible to adapt the quality profile to a possibly new quality profile of the reference biologic.

The biosimilar quality profile would be matched and adapted through the comparability exercise to the most recent quality profile of the reference biologic (assuming a change happened for the reference biologic and the quality profiles were determined to be dissimilar at that stage).

This approach assures that especially during clinical phases the best possible foundation of a successful, and therefore, short clinical comparability exercise are set, where under certain circumstances extrapolation of indication could be envisaged.

The current biosimilar regulation extents to mAbs, but more complex biologics such as vaccines are not particularity addressed. The "A-Mab case study in Bioprocess Development" is a comprehensive example for the introduction of the concepts of ICH Q8 R2, including design space for complex bio-molecules (101).

---

[73] Design space:
The multidimensional combination and interaction of input variables (e.g., material attributes) and process parameters that have been demonstrated to provide assurance of quality. Working within the design space is not considered as a change. Movement out of the design space is considered to be a change and would normally initiate a regulatory post approval change process. Design space is proposed by the applicant and is subject to regulatory assessment and approval, from the INTERNATIONAL CONFERENCE ON HARMONISATION OF TECHNICAL
REQUIREMENTS FOR REGISTRATION OF PHARMACEUTICALS FOR HUMAN USE
ICH HARMONISED TRIPARTITE GUIDELINE
PHARMACEUTICAL DEVELOPMENT
Q8(R2)
Current Step 4 version
dated August 2009

The design spaces arise usually as multivariate interactions of critical process parameters and critical quality attributes. Therefore, where those interactions exist and can be used to control the limits of the design space that can be eroded/evaluated. For a complex bio-molecule such as mAbs, the following critical quality attribute parameters are of major interest:

1. Aggregation
2. Glycosylation (galactosylation, fructosylation, sialysation, mannosylation, etc)

Each process parameter is investigated for influence on the cited critical quality attributes. By splitting up process parameters into a sequential order based on their respective process steps a comprehensive overview of the relationship of process step, critical process parameters and critical quality attributes can be established.

A control strategy assures that the design space is not exceeded. This represents a science and risk based approach that provides a high degree of assurance that all product quality targets are met. The reference biologic, i.e. knowledge/understanding gained from its physical, chemical, biological, and microbiological properties or characteristics, will influence and shape the development of the biosimilar.

7.3.2 Small trials

Here the strategy is to have first time success, and therefore, limited trials for the biosimilar developer.

- **The foundation for the clinical phase is to match the purity profiles and to justify impurity profile differences.**
- **To adapt throughout the development program the quality profile of the biosimilar to the reference biologic, which requires ongoing characterisation/monitoring.**
- To limit clinical trials, combined PK +PD studies should have been carried out as part of the phase I trials.
- Phase II is not required and phase III trials can be done by one pivotal (lot vs. lot) double blinded randomized and sufficiently powered study, where the endpoint holds surrogate markers.
- It can be of benefit to carry out a 4 arm clinical trial design where in one study the most distinct quality profiles of the reference biologic and the biosimilar will be evaluated for clinical comparability.

7.3.3 Extrapolation of Indications

Concerns have been expressed about extrapolation of efficacy and safety data, i.e. in indications that are approved for the reference biological but for which the biosimilar has not been investigated.

Biosimilars, in contrast to small molecule chemical generics, cannot automatically claim all indications of the reference product and any extrapolation of data requires sound scientific justification.

For extrapolation of data to be considered, the following requirements need to be fulfilled:

1. Similarity with the reference product must be demonstrated, based on the approach to mimic the quality profile of the reference biologic, which sets the boundaries for the biosimilar. This should also include a similar ratio, identity and concentration of all isoforms of the reference biologic.
2. Assuming that clinical comparability is successfully demonstrated in a key indication, extrapolation of efficacy and safety data to other indication(s) of the reference product may be possible, based on the condition that the **mechanism of action involved in the extrapolated indications are the same and well understood for every major isoform present within the reference biologic** to assure similar efficacy.
3. The safety profile of the biosimilar must have been properly extrapolated for the patient population in the indication, i.e. it requires investigation if the patient population for the key indication are comparable and adverse events are not expected to be different in the extrapolated indication, based on reasonable justification. That assumption subsequently requires an increased pharmacovigilance plan for this particular extrapolated indication.

Taken together, it is still arguable if the conditions can be met to assure that extrapolation to other indications of the reference biologic does not imply less reassurance as regards to efficacy and safety of the biosimilar. Even if all the aforementioned considerations are taken into account, additional safety, i.e. pharmacovigilance measures, should be taken, given the greater uncertainty with extrapolated biosimilars compared to normal biosimilars, where more experience was gained by the EMA.

Chapter 8.0 Integrated conclusion

During the course of this study evidence was presented and discussed related to the research questions. The fundamental question related to the heterogeneity and variation through the life cycle of the biosimilar and the reference biologic remains neglected, based on the results collected during this study, by the stakeholders, mainly because those issues where not addressed in any official regulatory guidance document.

As clear guidance from the EMA is lacking, with the only available guidance being that once a marketing authorization is issued it is accepted that the quality profiles of the reference biologic and the biosimilar will have different quality profiles. However, this statement already reveals that it is expected by the EMA that the quality profiles of the reference biologic and the biosimilar will diverge over the course of time.

There is no evidence available to show that changes of the quality profiles will not occur during the development phase of the biosimilar, prior to the marketing authorisation being issued for the biosimilar.

Deduced from the above, there is the need to further clarify the requirements for biosimilar developers, and possibly recommend a development strategy:

The QTPP (quality target product profile) established at the beginning of the biosimilar development program should be based on the quality, the safety and the efficacy profile from the reference, as a general aim.

Taking the development stages as a stepwise approach it is clear the first step should be characterizing the quality profile of the reference biologic to set the boundaries for the biosimilar quality profile; whereas, the boundaries of the quality profile should not exceed the heterogeneity and variability. The biosimilar requires such to extend the similarity to the reference biologic, as the reference biologic is similar to itself, when compared on a batch to batch basis, as a goal post for the quality profile. This means also that it is required to evaluate the purity and impurity profiles and their respective ratios, identity and quantities within a lot and between lots. At a later stage of the development program those aspect can be tested for their clinical impact and assuming that there is no detectable clinical difference due to the absence of similarity of the quality profiles, the biosimilar developer can justify different impurity profiles or different isoform rations. for example.

Following the stepwise approach further, the likelihood of conducting a successful clinical comparability exercise is presumably greater, in the case where the quality profile falls within the boundaries of the quality profile of the reference biologic.

The better the correlation of the molecular characteristics of the biosimilar and its correlation to quality attributes is understood, the more the biosimilar developer can shift the focus (of the development strategy) from mimicking the quality profile of the reference biologic, including its variability/heterogeneity, towards manufacturing a more consistent/purer biosimilar.

As a direct result of the lack of guidance on how to respond to changes in the quality profile of the reference biologic during the development phase of the biosimilar, the requirements for the later stages of the development of the biosimilar, namely the clinical trial based comparability exercise, are not clear. Biosimilarity requires similarity for the clinical aspects; a biosimilar should not be designed to be better at the clinical level than the reference biologic. That implies that non-inferiority trials should not be permitted for biosimilars. Only in the case where the quality profile of the biosimilar is shown to be within the boundaries of the quality profile of the reference biologic, it should be permitted to test only clinical comparability on a head to head (two arm design, with one arm for the reference biologic and the other arm for the biosimilar as well as a lot to lot clinical caparison in a sufficiently powered, double blinded, randomized study if possible) trial design. Otherwise in the case where the quality profile of the biosimilar is outside the boundaries of the reference biologic, multiple lots of the reference biologic may be tested against multiple lots of the biosimilar during the clinical comparability exercise.

Therefore, the biosimilar developer should be recommended to respond to any "sudden" change in the quality profile of the reference biologic in due time, assuring that the material used for the clinical trial comparability exercise is in fact within the boundaries of the quality profile of the reference biologic to limit the risks of failing the comparability exercise.

For particular scenarios, where the heterogeneity and variation of the reference biologic, even without a major change of the reference biologic, results in a detectable difference in the clinical comparability exercise, two possible causes need to be investigated.

1. The quality profile of the reference biologic and the biosimilar are not similar enough.
2. Different lots of the reference biologic fail to be comparable in clinical trials when compared to itself.

The reasons for such failure could be manifold, such as selection of the margin of tolerance and related to the selection of the clinical endpoint, or that the reference biologic is in its performance not consistent, i.e. the efficacy margins are low compared to the placebo arm, for example.

For these anticipated scenarios more guidance is needed, which will possibly only be available once the EMA is confronted with such a situation. One can only hope that that information will be made public for further analysis.

Further, the other impact heterogeneity and variation may have is on the possible extrapolation of indications. The mechanism of action not only is required to be identical, also the relative ratio (compared to each other), the identity and the quantity of the purities have a major impact on the extrapolation of indications. This should be firmly stated in EMA guidelines that extrapolation of indications can only be envisaged next to other conditions, in cases where the purity profile (isoform ratio towards each other, identify, quantity) of the biosimilar is within the boundaries of the purity profile of the reference biologic in addition to a comprehensive non-clinical data set showing the comparability of each of the different isoforms with regard to the extrapolated indication. This however is a burden a biosimilar company is unlikely to fulfil; therefore, a separate trial might be the more cost effective solution.

For the biosimilar developer, which is required to show comparability at a given moment in time towards the reference biologic, a design space approach would be the appropriate solution to be able to respond to sudden changes of the quality profile of the reference biologic.

However. as the dynamic of both quality profiles post marketing authorization will develop further and with no need for the respective products to follow each other's quality profiles any more, over time a situation might raise where the safety and efficacy profiles will be different and the initially shown comparability will be obsolete. It is assumed that EMA, as the authority having oversight over process changes will assure that such a situation will not happen; however, given the complexity of some biologics and the unpredictability of certain changes considered to be minor, it is arguable if such an approach, only based on regulatory control and oversight, is a good option.

Note:
It would be appreciated if more data would be made available from the industry and EMA. However, especially from the innovator as well the biosimilar industry, the willingness to share their experiences even for academic purposes was limited and not much detail was given. One can only regret such attitudes.

Bibliography or References

1. World Health Organization (2009) GUIDELINES ON EVALUATION OF SIMILAR BIOTHERAPEUTIC PRODUCTS (SBPs). , 19–23

2. European Medicines Agency (2012) Questions and answers on biosimilar medicines (similar biological medicinal products).

3. The European Parliment and the Council of the European Union (2004) DIRECTIVE 2001/83/EC OF THE EUROPEAN PARLIAMENT AND OF THE COUNCIL OF 6 NOVEMBER 2001 ON THE COMMUNITY CODE RELATING TO MEDICINAL PRODUCTS FOR HUMAN USE. , 67–128

4. Oxford Dictionaries (2013) Dictionaries, Oxford. *Oxford Dictionaries. online edition: Oxford University Press* [online] http://oxforddictionaries.com/.

5. EMEA (2008) EUROPEAN PUBLIC ASSESSMENT REPORT (EPAR) OMNITROPE EPAR summary for the public.

6. Weise, M., Bielsky, M.-C., De Smet, K., Ehmann, F., Ekman, N., Narayanan, G., Heim, H.-K., Heinonen, E., Ho, K., Thorpe, R., Vleminckx, C., Wadhwa, M., and Schneider, C. K. (2011) Biosimilars-why terminology matters. *Nature biotechnology* **29**, 690–693 [online] http://dx.doi.org/10.1038/nbt.1936 (Accessed February 11, 2013).

7. Sandoz (2011) Biosimilar Development. [online] http://www.sandoz-biosimilars.com/biosimilars/development.shtml (Accessed April 4, 2013).

8. European Medicines Agency (2011) Submission of comments on "Guideline on similar biological medicinal products containing monoclonal antibodies" (EMA / CHMP / BMWP / 403543 / 2010). , 1–73

9. Reichert, J. M. (2011) Next generation and biosimilar monoclonal antibodies: essential considerations towards regulatory acceptance in Europe. February 3-4, 2011, Freiburg, Germany. *mAbs* **3**, 223–240 [online] http://www.landesbioscience.com/journals/mabs/article/15475/ (Accessed February 7, 2013).

10. McCamish, M., and Woollett, G. (2011) Worldwide experience with biosimilar development. *mAbs* **3**, 209–217 [online] http://www.landesbioscience.com/journals/mabs/article/15005/ (Accessed January 29, 2013).

11. McCamish, M., and Woollett, G. (2011) Worldwide experience with biosimilar development. *mAbs* **3**, 209–217 [online]

http://www.landesbioscience.com/journals/mabs/article/15005/ (Accessed January 29, 2013).

12. Nick, C. (2006) in *Clinical Research Manual 2nd ed.* (Luscombe, D., and Stonier, P. D., eds.) pp. S21.1–S21.12, Euromed Communications, Haslemere, England

13. Hou, J. J. C., Codamo, J., Pilbrough, W., Hughes, B., Gray, P. P., and Munro, T. P. (2011) New frontiers in cell line development: challenges for biosimilars. *Journal of Chemical Technology & Biotechnology* **86**, 895–904 [online] http://doi.wiley.com/10.1002/jctb.2574 (Accessed February 11, 2013).

14. Schellekens, H. (2009) Biosimilar therapeutics—what do we need to consider? *NDT plus* **2**, i27–i36 [online] http://www.pubmedcentral.nih.gov/articlerender.fcgi?artid=2638545&tool=pmcentrez&rendertype=abstract (Accessed February 10, 2013).

15. Entine, J. (2012) FDA Balances Costs, Patient Safety in the Biologics and Personalized Medicine Revolution. *Forbes* [online] http://www.forbes.com/sites/jonentine/2012/07/23/fda-balances-costs-patient-safety-in-the-biologics-and-personalized-medicine-revolution-will-it-get-it-right-or-damage-the-miracle-industry/2/.

16. Parexel (2012) Biosimilar and biogenerics pharmaceutical development. [online] http://www.parexel.com/services-and-capabilities/consulting/biosimilars/.

17. Carroll, J. (2010) Blockbuster antibodies now face quick onslaught of biosimilars. [online] http://www.fiercebiotech.com/story/blockbuster-antibodies-now-face-quick-onslaught-biosimilars/2010-11-30.

18. MacDonald, R. (2012) Biologic copy-cat drug development is rising, bringing pharmaceutical companies together. *Medill Reports Chicago* [online] http://news.medill.northwestern.edu/chicago/news.aspx?id=205418.

19. Roger, S. D. (2010) Biosimilars : current status and future directions. **10**, 1–8

20. EUROPEAN GENERIC MEDICINES ASSOCIATION (2011) *FAQs about Biosimilar Medicines*, [online] http://www.egagenerics.com/images/factsheet/EGA_factsheet_07.pdf.

21. Kálmán-Szekeres, Z., Olajos, M., and Ganzler, K. (2012) Analytical aspects of biosimilarity issues of protein drugs. *Journal of pharmaceutical and biomedical analysis* **69**, 185–95 [online] http://www.ncbi.nlm.nih.gov/pubmed/22633839 (Accessed February 11, 2013).

22. Sandoz (2009) An important step for us, a giant leap for mankind: Biosimilars by Sandoz. [online] http://www.sandoz-biosimilars.ch/pdf/Biosimilar-Broschuere_engl.pdf.

23. Xia, Z., Cai, X., and Cao, P. (2012) Monoclonal antibody: the corner stone of modern biotherapeutics. *Yao Xue Xue Bao* **47**, 1275–80

24. GABI (2013) Biologicals boom. *GABI online* [online] http://gabionline.net/Biosimilars/Research/Biologicals-boom?utm_source=GONL&utm_campaign=3f2a27df68-GONL+V13B15&utm_medium=email (Accessed March 11, 2013).

25. Pisani, J., and Bonduelle, Y. *OPPORTUNITIES AND BARRIERS IN THE BIOSIMILAR MARKET: EVOLUTION OR REVOLUTION FOR GENERICS COMPANIES?*,

26. Declerck, P. (2013) Biosimilar monoclonal antibodies: a science-based regulatory challenge. *Expert Opinion on Biological Therapy* **13**, 153–156 [online] http://informahealthcare.com/doi/abs/10.1517/14712598.2012.758710 (Accessed February 10, 2013).

27. GABI (2011) Patent filing and market exclusivity. *GABI online* [online] http://www.gabionline.net/Policies-Legislation/Patent-filing-and-market-exclusivity.

28. Dunn, M. (2011) Timing of patent filing and market exclusivity. *Nature Reviews Drug Discovery* **10**, 487–8 [online] http://www.ncbi.nlm.nih.gov/pubmed/21720399 (Accessed February 10, 2013).

29. FDA (2009) Small Business Assistance: Frequently Asked Questions on the Patent Term Restoration Program. [online] http://www.fda.gov/Drugs/DevelopmentApprovalProcess/SmallBusinessAssistance/ucm069959.htm.

30. EMA (2013) Multidisciplinary: Biosimilar. [online] http://www.ema.europa.eu/ema/index.jsp?curl=pages/regulation/general/general_content_000408.jsp&mid=WC0b01ac058002958c (Accessed March 11, 2013).

31. National Center for Biotechnology Information (2013) PubMed. [online] http://www.ncbi.nlm.nih.gov/pubmed/ (Accessed March 11, 2013).

32. GABI (2013) GABI online. *GABI online* [online] http://www.gabionline.net/ (Accessed March 11, 2013).

33. Google (2013) Google. [online] https://www.google.com/ (Accessed March 11, 2013).

34. European Medicines Agency CHMP: Overview. [online] http://www.emea.europa.eu/ema/index.jsp?curl=pages/about_us/general/general_content_000095.jsp&mid=WC0b01ac0580028c7a (Accessed February 19, 2013).

35. European Commission The Centralised Procedure. [online] http://ec.europa.eu/health/authorisation-procedures-centralised_en.htm (Accessed February 19, 2013).

36. European Medicines Agency Multidisciplinary: Biosimilar. [online] http://www.ema.europa.eu/ema/index.jsp?curl=pages/regulation/general/general_content_000408.jsp&mid=WC0b01ac058002958c (Accessed February 19, 2013).

37. Schellekens, H. (2004) When biotech proteins go off-patent. *Trends in biotechnology* **22**, 406–10 [online] http://www.ncbi.nlm.nih.gov/pubmed/15283985 (Accessed January 29, 2013).

38. Schellekens, H. (2005) Follow-on biologics: challenges of the "next generation". *Nephrology, dialysis, transplantation* **20**, iv31–36 [online] http://www.ncbi.nlm.nih.gov/pubmed/15827057 (Accessed February 11, 2013).

39. Pray, L. (2008) DNA Replication and Causes of Mutation. *Nature Education* **1**

40. Locatelli, F., and Roger, S. (2006) Comparative testing and pharmacovigilance of biosimilars. *Nephrology, dialysis, transplantation* **21**, v13–6 [online] http://www.ncbi.nlm.nih.gov/pubmed/16959789 (Accessed February 11, 2013).

41. Kuhlmann, M., and Covic, A. (2006) The protein science of biosimilars. *Nephrology, dialysis, transplantation* **21**, v4–8 [online] http://www.ncbi.nlm.nih.gov/pubmed/16959791 (Accessed February 11, 2013).

42. Liang, B. A. (2006) ARTICLE REGULATING FOLLOW-ON BIOLOGICS. *Harvard Journal on Legislation* **44**, 363–460

43. Pilbrough, W., Munro, T. P., and Gray, P. (2009) Intraclonal protein expression heterogeneity in recombinant CHO cells. *PloS one* **4**, e8432 [online] http://www.pubmedcentral.nih.gov/articlerender.fcgi?artid=2793030&tool=pmcentrez&rendertype=abstract (Accessed February 11, 2013).

44. Kresse, G.-B. (2009) Biosimilars--science, status, and strategic perspective. *European journal of pharmaceutics and biopharmaceutics* **72**, 479–86 [online] http://www.ncbi.nlm.nih.gov/pubmed/19286455 (Accessed January 29, 2013).

45. Ahmed, I., Kaspar, B., and Sharma, U. (2012) Biosimilars: impact of biologic product life cycle and European experience on the regulatory trajectory in the United States. *Clinical therapeutics* **34**, 400–19 [online] http://www.ncbi.nlm.nih.gov/pubmed/22244050 (Accessed February 11, 2013).

46. Depp, V., Alikhani, A., Grammer, V., and Lele, B. S. (2009) Native protein-initiated ATRP: a viable and potentially superior alternative to PEGylation for stabilizing biologics. *Acta biomaterialia* **5**, 560–9 [online] http://www.ncbi.nlm.nih.gov/pubmed/18804423 (Accessed February 11, 2013).

47. Grabowski, H. (2008) Follow-on biologics: data exclusivity and the balance between innovation and competition. *Nature Reviews Drug Discovery* **7**, 479–488

48. European Medicines Agency (2010) Guideline on similar biological medicinal products containing monoclonal antibodies. , 1–13

49. EMEA (2005) GUIDELINE ON SIMILAR BIOLOGICAL MEDICINAL PRODUCTS.

50. EMEA (2003) GUIDELINE ON COMPARABILITY OF MEDICINAL PRODUCTS CONTAINING BIOTECHNOLOGY-DERIVED PROTEINS AS ACTIVE SUBSTANCE: QUALITY ISSUES. , 1–11

51. EMEA (2003) GUIDELINE ON COMPARABILITY OF MEDICINAL PRODUCTS CONTAINING BIOTECHNOLOGY-DERIVED PROTEINS AS ACTIVE SUBSTANCE NON-CLINICAL AND CLINICAL ISSUES.

52. European Medicines Agency (2011) Concept paper on the revision of the guideline on similar biological medicinal product.

53. EMEA (2006) GUIDELINE ON SIMILAR BIOLOGICAL MEDICINAL PRODUCTS CONTAINING BIOTECHNOLOGY-DERIVED PROTEINS AS ACTIVE SUBSTANCE: QUALITY ISSUES.

54. European Medicines Agency (2012) Guideline on similar biological medicinal products containing biotechnology-derived proteins as active substance: quality issues (revision 1).

55. EMEA (2006) GUIDELINE ON SIMILAR BIOLOGICAL MEDICINAL PRODUCTS CONTAINING BIOTECHNOLOGY-DERIVED PROTEINS AS ACTIVE SUBSTANCE: NON-CLINICAL AND CLINICAL ISSUES.

56. European Medicines Agency (2011) Concept paper on the revision of the guideline on similar biological medicinal products containing biotechnology- derived proteins as active substance : non-clinical and clinical issues.

57. FDA (2012) FDA issues draft guidance on biosimilar product development. [online] http://www.fda.gov/NewsEvents/Newsroom/PressAnnouncements/ucm291232.htm (Accessed February 19, 2013).

58. GABI (2012) Overview of research on US regulatory issues surrounding biosimilars in 2012. *GABI online* [online] http://www.gabionline.net/Biosimilars/Research/Overview-of-research-on-US-regulatory-issues-surrounding-biosimilars-in-2012 (Accessed February 19, 2013).

59. PhRMA Guidelines for the Quality, Safety and Efficacy Assurance of Follow-on Biologics.

60. Arato, T. (2011) Recent Regulations of Biosimilars in Japan.

61. Tessensohn, J. A., and Yamamoto, S. (2011) Patent term extensions for biologic innovators in Japan. *Nature Biotechnology* **29**, 34–37

62. Health Canada (2010) Guidance For Sponsors: Information and Submission Requirements for Subsequent Entry Biologics (SEBs). [online] http://www.hc-sc.gc.ca/dhp-mps/brgtherap/applic-demande/guides/seb-pbu/seb-pbu_2010-eng.php (Accessed May 13, 2013).

63. Randolph, J. J. (2009) A Guide to Writing the Dissertation Literature Review. *Practical Assessment, Research and Evaluation* **14** [online] http://pareonline.net/pdf/v14n13.pdf.

64. Chandrasekhar, R. (2000) How to Write a Thesis: A Working Guide. [online] http://ciips.ee.uwa.edu.au/pub/HowToWriteAThesis.pdf.

65. Leung, W.-C. (2001) How to design a questionnaire. *Student BMJ* **9**, 187–189 [online] http://www.cochrane.es/files/Recursos/How_to_design_a_questionnaire.pdf.

66. Local Government Data Unit (2009) A guide to... ...Questionnaire Design. [online] http://www.dataunitwales.gov.uk/SharedFiles/Download.aspx?pageid=79&fileid=447&mid=459.

67. EMA (2012) List of participants: Closed Workshop on biosimilar monoclonal antibodies and immunogenicity of monoclonal antibodies Monday 24P th P October 2011. [online] http://www.ema.europa.eu/docs/en_GB/document_library/Other/2012/06/WC500128738.pdf.

68. Danzer, K., and Curries, L. D. (1998) Guidelines for Calibration in Analytical Chemistry: Part 1. Fundamentals and Single Component Calibration. *Pure and Applied Chemistry* **70**, 993–1014

69. European Medicines Agency (2012) Overview of comments received on ' Guideline on similar biological medicinal products containing monoclonal antibodies ' (EMA / CHMP / BMWP / 403543 / 2010).

70. Reichert, J. M., and Beck, A. (2009) European Medicines Agency workshop on biosimilar monoclonal antibodies: July 2, 2009, London, UK. *mAbs* **1**, 394–416 [online] http://www.landesbioscience.com/journals/mabs/article/9630/.

71. Schellekens, H., and Moors, E. (2010) Clinical comparability and European biosimilar regulations. *Nature biotechnology* **28**, 28–31 [online] http://www.ncbi.nlm.nih.gov/pubmed/20062035 (Accessed February 7, 2013).

72. Declerck, P. J., Darendeliler, F., Góth, M., Kolouskova, S., Micle, I., Noordam, C., Peterkova, V., Volevodz, N. N., Zapletalová, J., and Ranke, M. B. (2010) Biosimilars: controversies as illustrated by rhGH. *Current medical research and opinion* **26**, 1219–29 [online] http://www.ncbi.nlm.nih.gov/pubmed/20302553 (Accessed February 10, 2013).

73. Nick, C. (2011) How can the biosimilar concept be applied to monoclonals? *Regulatory Rapporteur*, 11–16

74. Dowlat, H. A. (2010) The current status of biosimilar biologics – Part 2: Practical considerations in international development through European lessons learned. *Regulatory Rapporteur* **7**

75. European Commission (2012) The Rules Governing Medicinal Products in the European Union Volume 4 EU guidelines for Good Manufacturing Practice for Medicinal Products for Human and Veterinary Use.

76. Crombie, I. K., and Davies, H. T. O. (2009) What is meta-analysis? , 1–8

77. Barosi, G., Bosi, A., Abbracchio, M. P., Danesi, R., Genazzani, A., Corradini, P., Pane, F., and Tura, S. (2011) Key concepts and critical issues on epoetin and filgrastim biosimilars. A position paper from the Italian Society of Hematology, Italian Society of Experimental Hematology, and Italian Group for Bone Marrow Transplantation. *Haematologica* **96**, 937–42 [online] http://www.haematologica.org/content/96/7/937 (Accessed February 10, 2013).

78. GABI (2013) Positive phase I data for infliximab biosimilar. *GABI online* [online] http://gabionline.net/Biosimilars/Research/Positive-phase-I-data-for-infliximab-biosimilar/(highlight)/phase III (Accessed March 12, 2013).

79. GABI (2012) Samsung halts biosimilar rituximab development. *GABI online2* [online] http://gabionline.net/Biosimilars/News/Samsung-halts-biosimilar-rituximab-development/(highlight)/phase III (Accessed March 12, 2013).

80. GABI (2012) Teva halts phase III biosimilar rituximab trial. *GABI online* [online] http://gabionline.net/Biosimilars/News/Teva-halts-phase-III-biosimilar-rituximab-trial/(highlight)/phase III (Accessed April 2, 2013).

81. Fox, A. (2010) Biosimilar medicines--new challenges for a new class of medicine. *Journal of Biopharmaceutical Statistics* **20**, 3–9 [online] http://www.ncbi.nlm.nih.gov/pubmed/20077245 (Accessed February 7, 2013).

82. GABI (2012) Boehringer Ingelheim starts biosimilar rituximab trial. *GABI online* [online] http://gabionline.net/Biosimilars/Research/Boehringer-Ingelheim-starts-biosimilar-rituximab-trial/(highlight)/phase III (Accessed May 3, 2013).

83. GABI (2013) Finox submits r-FSH biosimilar application to EMA. *GABI online* [online] http://gabionline.net/Biosimilars/News/Finox-submits-r-FSH-biosimilar-application-to-EMA/(highlight)/phase III (Accessed May 3, 2013).

84. GABI (2012) Positive results from phase III study with biosimilar human insulin. *GABI online* [online] http://gabionline.net/Biosimilars/Research/Positive-results-from-phase-III-study-with-biosimilar-human-insulin/(highlight)/phase III (Accessed May 3, 2013).

85. GABI (2012) Sandoz starts phase III US trial for biosimilar epoetin alfa. *GABI online* [online] http://gabionline.net/Biosimilars/Research/Sandoz-starts-phase-III-US-trial-for-biosimilar-epoetin-alfa/(highlight)/phase III (Accessed May 3, 2013).

86. GABI (2012) Biosimilar infliximab equivalence proven in phase III trial. *GABI online* [online] http://gabionline.net/Biosimilars/Research/Biosimilar-infliximab-equivalence-proven-in-phase-III-trial/(highlight)/phase III (Accessed February 3, 2013).

87. GABI (2012) Successful phase III trial of biosimilar r-FSH. *GABI online* [online] http://gabionline.net/Biosimilars/News/Successful-phase-III-trial-of-biosimilar-r-FSH/(highlight)/phase III (Accessed May 3, 2013).

88. GABI (2012) Successful completion of biosimilar infliximab programme. *GABI online* [online] http://gabionline.net/Biosimilars/News/Successful-completion-of-biosimilar-infliximab-programme/(highlight)/phase III (Accessed May 3, 2013).

89. GABI (2012) Hospira starts phase III programme for biosimilar erythropoietin. *GABI online* [online] http://gabionline.net/Biosimilars/News/Hospira-starts-phase-III-programme-for-biosimilar-erythropoietin/(highlight)/phase III (Accessed May 3, 2013).

90. GABI (2010) Phase III study of a new biosimilar filgrastim product (Zarzio) published. *GABI online* [online] http://gabionline.net/Biosimilars/Research/Phase-III-study-of-a-new-biosimilar-filgrastim-product-Zarzio-published/(highlight)/phase III (Accessed May 3, 2013).

91. Waller, C. F., Semiglazov, V. F., Tjulandin, S., Bentsion, D., Chan, S., and Challand, R. (2010) A phase III randomized equivalence study of biosimilar filgrastim versus Amgen filgrastim in patients receiving myelosuppressive chemotherapy for breast cancer. *Onkologie* **33**, 504–511

92. Romer, T., Peter, F., Saenger, P., Starzyk, J., Koehler, B., Korman, E., Walczak, M., Wasik, R., Ginalska-Malinowska, M., Solyom, E., and Berghout, A. (2007) Efficacy and safety of a new ready-to-use recombinant human growth hormone solution. *Journal of endocrinological investigation* **30**, 78–89

93. Engert, A., Del Giglio, A., Bias, P., Lubenau, H., Gatzemeier, U., and Heigener, D. (2009) Incidence of febrile neutropenia and myelotoxicity of chemotherapy: a meta-analysis of biosimilar G-CSF studies in breast cancer, lung cancer, and non-Hodgkin's lymphoma. *Onkologie* **10**, 599–604

94. López-Siguero, J., Pérez, M. V. B., Balser, S., and Khan-Boluki, J. (2011) Long-term Safety and Efficacy of the Recombinant Human Growth Hormone Omnitrope® in the Treatment of Spanish Growth Hormone Deficient Children: Results of a Phase III Study. *Advances in Therapy* **28**, 879–893

95. Tzekova, V., Mihaylov, G., Elezovic, I., and Koytchev, R. (2009) Therapeutic effects of epoetin zeta in the treatment of chemotherapy-induced anaemia. *Current Medical Research and Opinion* **25**, 89–97

96. Engert, A., Griskevicius, L., Zyuzgin, Y., Lubenau, H., and Del Giglio, A. (2009) XM02, the first granulocyte colony-stimulating factor biosimilar, is safe and effective in reducing the duration of severe neutropenia and incidence of febrile neutropenia in patients with non-Hodgkin lymphoma receiving chemotherapy. *Leukemia and Lymphoma* **50**, 374–379

97. Gatzemeier, U., Ciuleanu, T., Dediu, M., Ganea-Motan, E., Lubenau, H., and Giglio, A. D. (2009) XM02, the First Biosimilar G-CSF, is Safe and Effective in Reducing the Duration of Severe Neutropenia and Incidence of Febrile Neutropenia in Patients with Small Cell or Non-small Cell Lung Cancer Receiving Platinum-Based Chemotherapy. *Journal of Thoracic Oncology* **4**

98. European Medicines Agency (2011) Concept paper on the revision of the guideline on similar biological medicinal products containing biotechnologyderived proteins as active substance: non-clinical and clinical issues.

99. GABI (2013) Quality, similarity and safety of biosimilars. *GABI online* [online] http://gabionline.net/Biosimilars/Research/Quality-similarity-and-safety-of-biosimilars?utm_source=GONL&utm_campaign=81b75ff64f-GONL+V13D12&utm_medium=email (Accessed May 2, 2013).

100. GABI (2013) Celltrion stops phase III trial of biosimilar rituximab. *GABI online* [online] http://gabionline.net/Biosimilars/News/Celltrion-stops-phase-III-trial-of-biosimilar-rituximab?utm_source=GONL&utm_campaign=f84d1b631e-GONL+V13D29&utm_medium=email (Accessed April 2, 2013).

101. CMC Biotech Working Group (2009) *A-Mab case study in Bioprocess Development,*

102. Czajkowsky, D. M., Hu, J., Shao, Z., and Pleas, R. J. (2012) Fc-fusion proteins: new developments and future perspectives. *EMBO Mol Med* **4**, 1015–1028

List of Appendices

Appendix: 1

Questions and comments submitted to the EMA for the "Guideline on similar biological medicinal products containing monoclonal antibodies Draft".

Appendix: 2

Questions and comments submitted to the EMA for the "Guideline on similar biological medicinal products containing biotechnology-derived proteins as active substance: quality issues (revision 1)"

Appendix: 3

Questionnaire 1.1

Appendix: 4

Questionnaire 1.2

Appendix: 5

Answers from the EMA to the "Questions and comments submitted to the EMA for the "Guideline on similar biological medicinal products containing monoclonal antibodies Draft".

Appendix: 1

Questions and comments submitted to the EMA for the "Guideline on similar biological medicinal products containing monoclonal antibodies Draft".

Submission of comments on: <30.05.11>

'<Guideline on similar biological medicinal products containing monoclonal antibodies Draft>' (EMA.europa.eu/docs/en_GB/document_library/Scientific_guideline/2010/11/WC500099361.pdf)

Comments from:

Name of organisation or individual
Malik Osmane, Diplom Biologe and student of "Pharmaceutical Quality Assurance", MSc. program

Please do not hesitate to contact me for further discussion on my comments.

1. General comments

Stakeholder number *(To be completed by the Agency)*	General comment (if any)	Outcome (if applicable) *(To be completed by the Agency)*
	This guidance (Guideline on similar biological medicinal products containing monoclonal antibodies) is surely an important step forward with regard to facilitating patent access to safe and effective but also less costly medicinal products. I further think that in relation to already existing product specific guidance documents, this new guidance document is the most progressive due to its wider scope and forward viewing emphasis for example with regard to the introduction of the concept of Extrapolation of indication for a Biosimilar The explicit mentioning of and encouraged usage of "Meta Analysis" in this guide However as in any progressive document, there are few areas which require further clarification. Although these areas are not specifically linked to mAbs and moreover to Biosimilars in a more general way, this guide uses some key terminology and concepts such as: 1. Extrapolation of Indications 2. Meta Analysis 3. Conventional Equivalence Margin , which require further explanation. I feel it would be appropriate either to clarify those terms and concepts as part of this guide or to address them in a separate guide, applicable for all Biosimilars taking into account the heterogeneity and the nature of mAbs (and other Biosimilars). The EMA guide EMEA/CHMP/BWP/49348/2005 and subdivides variability in: i) Product variability ii) Process related variability. Those two variability types form what is called the "Quality Profile" of the Biosimilar. However these types of variability are relevant for all biotechnological derived medicinal products, be it the Reference Medicinal Product or be it the Biosimilar.	
	One particular issue is in my opinion the variability of the Reference Medicinal Product itself. 1.) Therefore the variability of the Reference Medicinal Product	

Stakeholder number (To be completed by the Agency)	General comment (if any)	Outcome (if applicable) (To be completed by the Agency)
	itself (as a variable) leads to questions on how to fulfill requirements for the Biosimilar i.e. to demonstrate comparable clinical efficacy and safety versus the Reference Medicinal Product. Or differently expressed: 2.) When requiring any direct comparison, be it in a non-clinical or clinical setting, batch to batch variation of the Reference Medicinal Product (due to the glycosylation heterogeneity of mAbs or process impurities associated with the production of mAbs i.e. the full quality profile of the Reference Medicinal Product) and the quality profile of the Biosimilar need to be taken into account when designing studies and analyzing data. I call this train of thoughts in this submission the "overlaying issue of product heterogeneity and variation [(1)]".	
	This "overlaying issue of product heterogeneity and variation [(1)]" is founded on and also (partially) reflected in the following documents: *1.) "Concept paper on the revision of the guideline on similar biological medicinal products containing biotechnology-derived proteins as active substance: quality issues".* *Section 3 (Problem statement)* Which takes into account the life cycle of the Biosimilar and variations of the process, impacting the variability (Quality Profile) of the Biosimilar. This is of relevance for any Biological Medicinal Product, be it a Reference Medicinal Product or Biosimilar. This "overlaying issue of product heterogeneity and variation [(1)]", appears also to be justified taken into account the exiting guidance: *2.) Doc. Ref. EMEA/CHMP/BMWP/101695/2006* *GUIDELINE ON COMPARABILITY OF BIOTECHNOLOGY-DERIVED MEDICINAL PRODUCTS AFTER A CHANGE IN THE MANUFACTURING PROCESS NON-CLINICAL AND CLINICAL ISSUES* *"Demonstration of comparability is a sequential process, beginning with quality studies (limited or comprehensive) and supported, as necessary, by non-clinical and/or clinical studies.* *This guideline will address the requirements for non-clinical and/or clinical bridging studies to demonstrate that the modification has no impact on safety and efficacy.* *The selection of non-clinical versus clinical testing is product-driven, i.e. a model should be chosen that best detect clinically relevant differences with sufficient accuracy"* *3.) Doc Ref: EMEA/CHMP/BWP/49348/2005*	

Stakeholder number	General comment (if any)	Outcome (if applicable)
(To be completed by the Agency)		*(To be completed by the Agency)*
	GUIDELINE ON SIMILAR BIOLOGICAL MEDICINAL PRODUCTS CONTAINING BIOTECHNOLOGY-DERIVED PROTEINS AS ACTIVE SUBSTANCE: QUALITY ISSUES *"Consequently, an extensive comparability exercise will be required to* *demonstrate that the biosimilar and reference products have similar profiles in terms of quality, safety* *and efficacy.* *Based on the comparability approach and when supported by sufficiently sensitive analytical systems,* *the demonstration of comparability at the quality level may connect the biosimilar product to the* *nonclinical and clinical data previously generated with the reference product"* *"It is recognised that each medicinal product is defined by the molecular composition of the active* *substance resulting from its process, which may introduce its own process related impurities.* *Consequently, the biosimilar product is defined by the following two sets of characteristics: i) related* *to the characteristics of the molecule (including product related substances/impurities), and ii) related* *to its process"* *4.) CHMP/437/04* *GUIDELINE ON SIMILAR BIOLOGICAL MEDICINAL PRODUCTS* *"Biological medicinal products are usually more difficult to characterize than chemically derived* *medicinal products. In addition, there is a spectrum of molecular complexity among the various* *products (recombinant DNA, blood or plasma-derived, immunologicals, gene and cell-therapy, etc.).* *Moreover, parameters such as the three-dimensional structure, the amount of acido-basic variants or* *post-translational modifications such as the glycosylation profile can be significantly altered by* *changes, which may initially be considered to be 'minor' in the manufacturing process. Thus, the* *safety/efficacy profile of these products is highly dependent on the robustness and the monitoring of* *quality aspects.* *Therefore:* *– The standard generic approach (demonstration of bioequivalence with a reference medicinal* *product by appropriate bioavailability studies) is normally applied to chemically derived* *medicinal products. Due to the complexity of biological/biotechnology-derived products the* *generic approach is scientifically not appropriate for these*	

Stakeholder number (To be completed by the Agency)	General comment (if any)	Outcome (if applicable) (To be completed by the Agency)
	products. The " similar biological medicinal products" approach, based on a comparability exercise, will then have to be followed"	

2. Specific comments on text

Line number(s) of the relevant text (e.g. Lines 20-23)	Stakeholder number (To be completed by the Agency)	Comment and rationale; proposed changes (If changes to the wording are suggested, they should be highlighted using 'track changes')	Outcome (To be completed by the Agency)
lines 135 to 138 of the draft guide.		130 *4.1. In vitro pharmacodynamic (PD) studies = step1* 131 In order to assess any difference in biological activity between the similar biological medicinal and the 132 reference medicinal product, data from a number of comparative *in vitro* studies, some of which may 133 already be available from quality-related assays, should be provided. Comment: Depending on assay sensitivity and methodology, comparative in vitro studies will always reveal differences in parameters cited in lines 135 to 138 of the draft guide. This is based on the reasonable assumption (based on available publications), that even the Reference Medicinal Product itself will vary between every batch, in for example the ratio glycosylated proteins and the level of glycosylation, refer also to [(1)]. Therefore the cited parameters will vary (slightly) for each batch of the Reference Medicinal Product or the Biosimilar, (again assuming sufficiently high assay sensitivity and methodology). Proposed change (if any): A statement at the beginning of section 4.0 should incorporate the following:	

Line number(s) of the relevant text *(e.g. Lines 20-23)*	Stakeholder number *(To be completed by the Agency)*	Comment and rationale; proposed changes *(If changes to the wording are suggested, they should be highlighted using 'track changes')*	Outcome *(To be completed by the Agency)*
		1.) "the Biosimilar requires to such an extend similarity to the Reference Medicinal Product, as the Reference Medicinal Product is similar to itself, when compared on a batch to batch basis, as the optimum. Deviation of this approach is generally acceptable, but requires justification and possibly additional clinical evaluation in order to assess associated risks" Further to guide the industry and to avoid unnecessary costs and disappointments, it should be stated that: 2.) "For non clinical setting it is advisable that the "Quality Profile", (refer to [(1)]) of the Biosimilar is within the levels of variation observed for the "Quality Profile" of the Reference Medicinal Product. Thoroughly determination of the "Quality Profile" of the Reference Medicinal Product can be used to set the margins of tolerance (for variation) for the Biosimilar "Quality Profile". Deviation of this approach is generally acceptable, but requires justification and possibly additional clinical evaluation in order to assess associated risks". By adopting this proposed approach, the level of comparability between the Reference Medicinal Product and the Biosimilar, using sound, measurable and justified parameters will reduce risks associated with clinical trials and will likely lead to (at least) a similar efficacy and safety profile of the Biosimilar in clinical trials compared to the Reference Medicinal Product	
Referring to lines: 209 to 212		209 The primary objective of the pharmacokinetic studies performed to support a Marketing Authorisation 210 Application (MAA) for a similar	

Line number(s) of the relevant text *(e.g. Lines 20-23)*	Stakeholder number *(To be completed by the Agency)*	Comment and rationale; proposed changes *(If changes to the wording are suggested, they should be highlighted using 'track changes')*	Outcome *(To be completed by the Agency)*
Referring to lines: 239 and 240 Referring to lines: 254 to 256		biological medicinal product is to show comparability in 211 pharmacokinetics of the biosimilar with the reference product in a sufficiently sensitive and 212 homogeneous population. 239 clearance and half-life should be determined and reported in a descriptive manner. If relevant 240 differences occur the assumption of similar PK might be seriously questioned. If such results are As a principle, any 254 widening of the conventional equivalence margin beyond 80-125% requires thorough justification, 255 including an estimation of potential impact on clinical efficacy and safety. This should be discussed with 256 regulatory authorities. Comment: For the clinical aspects the same point raised in (1) is valid. 1.) Assuming the Biosimilar needs to be as safe and effective as the Reference Medicinal Product as indicated in lines: 209 to 212 and that the clinical efficacy of the Biosimilar is required to be similar to the Reference Medicinal Product. 2.) The PK studies require to be similar compared to the Reference Medicinal Product. Referring to lines: 239 and 240 3.) Taken also lines: 254 to 256 into account, clearly the safety and efficacy comparability exercise is subject to tolerance margins. However this margin of tolerance is not clear and would require further elucidation. What is the justification for a "generally acceptable" margin with regard to	

Line number(s) of the relevant text *(e.g. Lines 20-23)*	**Stakeholder number** *(To be completed by the Agency)*	**Comment and rationale; proposed changes** *(If changes to the wording are suggested, they should be highlighted using 'track changes')*	**Outcome** *(To be completed by the Agency)*
		Biosimilars? Taken into consideration the following points: 1.) the problems associated with populations for trials for example in a cancer setting where sufficiently homogenous population might be rare and healthy volunteers cannot be used for ethical reasons either. 2.) cases where the "Quality Profile" of the Reference Medicinal Product exceeds already an efficacy margin of 80-125% when compared on a batch to batch basis in clinical trial to itself. There is also additional clarification required, from the innovator industries point of view, mainly on the following points: 1.) Taken the costs associated with clinical trials into account and the need to reduce healthcare costs, is it required to assess multiple batches of the Reference Medicinal Product when comparing it with the Biosimilar in a clinical trial setting, as a general question? 2.) As indicated in lines: 209 to 212 the expectation is that safety and efficacy are "comparable". A) Does this mean the acceptable tolerance margin for the Biosimilar in a clinical efficacy trial is required to be within the variation of endpoints of <u>multiple batches</u> of the Reference Medicinal Product and then the conventional equivalence margin of 80%-125% for the Biosimilar is appicable? Or	

Line number(s) of the relevant text *(e.g. Lines 20-23)*	Stakeholder number *(To be completed by the Agency)*	Comment and rationale; proposed changes *(If changes to the wording are suggested, they should be highlighted using 'track changes')*	Outcome *(To be completed by the Agency)*
		B) Does this mean the acceptable tolerance margin for the Biosimilar in a clinical efficacy trial is based on one batch the Reference Medicinal Product and then the conventional equivalence margin of 80-125% for the Biosimilar is applicable? Proposed change (if any): None, i.e. clarification required.	
		424 6. Extrapolation of Indications 425 Extrapolation of clinical efficacy and safety data to other indications of the reference mAb, not 426 specifically studied during the clinical development of the biosimilar mAb, is possible based on the 427 overall evidence of biosimilarity provided from the comparability exercise and with adequate 428 justification. Comment: Referring to lines 424 to 428, what would the justification be? As the mechanism of action of the Reference Medicinal Product could be different from indication to indication, is it justified to allow "Extrapolation of Indications" without showing the efficacy for this particular indication for the Biosimilar? For example for some indication an mAb might bind to a cell surface, whereas in other indications the primary action is to bind to soluble antigen. The subsequent clearance and inactivation mechanism might be very different. The EMA wrote on this topic the following: 1.) *2 EMA/CHMP/BMWP/86289/2010* *3 Committee for Medicinal Products for Human Use (CHMP)*	

Line number(s) of the relevant text *(e.g. Lines 20-23)*	Stakeholder number *(To be completed by the Agency)*	Comment and rationale; proposed changes *(If changes to the wording are suggested, they should be highlighted using 'track changes')*	Outcome *(To be completed by the Agency)*
		4 Guideline on immunogenicity assessment of monoclonal *5 antibodies intended for in vivo clinical use.* *6 Draft* *"261 MAbs exert their action by various mechanisms ranging from simple binding to antigen, which alone* *262 mediates the clinical effect, to binding antigen and mediating one or more immunobiological* *263 mechanisms which combine to produce the overall clinical response. Therefore, although simple* *264 binding may seem to be the only mechanism operating to achieve clinical efficacy, other effects may* *265 also play a role in this."* *And* *"the mode of action of the mAb (e.g., lytic, apoptotic), and especially the nature of* *331 the target molecule (e.g., immunosilencing, immunostimulating), needs to be adequately characterized* *332 and comprehensively investigated"* Differences in the glycosylation pattern for example between the Reference Medicinal Product and Biosimilar could be for one mechanism of action negligible, whereas for another mechanism of action very important. The mechanism of action can be different from indication to indication. As the mechanism of action can be influenced by the glycosylation pattern, which can be influenced by the "Quality Profile" (refer to "overlaying issue of product heterogeneity and variation [(1)]"); an important consequence of this train of thoughts is that new clinical efficacy data would be required for a new indication in order to ascertain that efficacy of the Biosimilar is comparable to the efficacy of	

Line number(s) of the relevant text *(e.g. Lines 20-23)*	Stakeholder number *(To be completed by the Agency)*	Comment and rationale; proposed changes *(If changes to the wording are suggested, they should be highlighted using 'track changes')*	Outcome *(To be completed by the Agency)*
		the Reference Medicinal Product for this new indication (assuming the case of a different mechanism of action). Proposed change (if any): None, i.e. clarification required.	

Appendix: 2

Questions and comments submitted to the EMA for the "Guideline on similar biological medicinal products containing biotechnology-derived proteins as active substance: quality issues (revision 1)"

<28.11.2012>

Submission of comments on '< Guideline on similar biological medicinal products containing biotechnology-derived proteins as active substance: quality issues (revision 1) >' (EMA/CHMP/BWP/247713/2012)

Comments from:

Name of organisation or individual

Malik Osmane

Please note that these comments and the identity of the sender will be published unless a specific justified objection is received.

When completed, this form should be sent to the European Medicines Agency electronically, in Word format (not PDF).

1. General comments

Stakeholder number *(To be completed by the Agency)*	General comment (if any)	Outcome (if applicable) *(To be completed by the Agency)*
	Letter to EMA This biosimilar guide from EMA addresses an array of questions and is again great step forward in biosimilar regulation. In the proposed guide EMA acknowledges some comments from previous discussions and gave meaningful guidance, as EMA recommends to evaluate multiple lots of the reference biologic at various stages of its shelf life, as a basis to establish a QTPP for the biosimilar. This precision/clarification is very much appreciated. I am delighted that this precision corresponds to some comments which were made in the past. *Stakeholder 13 comment:* *…"the Biosimilar requires to such an extend similarity to the Reference Medicinal Product, as the Reference Medicinal Product is similar to itself, when compared on a batch to batch basis, as the optimum. Deviation of this approach is generally acceptable, but requires justification and possibly additional clinical evaluation in order to assess associated risks"…* *"For non clinical setting it is advisable that the "Quality Profile", (refer to (1)) of the Biosimilar is within the levels of variation observed for the "Quality Profile" of the Reference Medicinal Product. Thoroughly determination of*	

Stakeholder number *(To be completed by the Agency)*	General comment (if any)	Outcome (if applicable) *(To be completed by the Agency)*
	the "Quality Profile" of the Reference Medicinal Product can be used to set the margins of tolerance (for variation) for the Biosimilar "Quality Profile". Deviation of this approach is generally acceptable, but requires justification and possibly additional clinical evaluation in order to assess associated risks" *EMA response:* *..."These topics are addressed in the " Guideline on similar biological medicinal products containing biotechnology-derived proteins as active substance: quality issues " which is currently under revision"...* *EMA/205886/2012, Page 154/422, stakeholder 13.* As in any new document there is room for further clarification improvement. I would like to bring to your attention three points, where some further improvement could be made or some clarification would be required. The aim is to contribute to patients health with useful and stimulating input to this guide and while appreciating EMAs leading role in tackling the challenges associated with biosimilars. My best regards Malik Osmane (Diplom Biologe) --- **3rd Comment:** Here I would like to focus on the understanding of the heterogeneity/variation of the two	

Stakeholder number *(To be completed by the Agency)*	General comment (if any)	Outcome (if applicable) *(To be completed by the Agency)*
	biologics (reference biologic and biosimilar) and their combined impact on the regulatory situation. 3.1 First, I would like to ask EMA to kindly clarify some comments on the quality profile and to highlight if these points were incorporated into this guide as announced: *Stàkeholder 13 comment:* *"...The EMA guide EMEA/CHMP/BWP/49 348/2005 and subdivides variability in:* *i) Product variability* *ii) Process related variability.* *Those two variability types form what is called the "Quality Profile" of the Biosimilar."* *EMA response:* *"The "quality profile" is these two sorts of variability PLUS the usual stand-alone physicochemical and biological characterisation PLUS a comprehensive comparability exercise. As such, the "quality programme" for a biosimilar is more extensive than that of a stand-alone."* *EMA/205886/2012, Page 29/422, Stakeholder 13* I consider variability/heterogeneity (derived from product/process) of the biosimilar as something which is to discover and to determine by	

Stakeholder number *(To be completed by the Agency)*	General comment (if any)	Outcome (if applicable) *(To be completed by the Agency)*
	characterization. The Q-profile is based on characterization data (physicochemical and biological). Characterization methods do have an error margins (as any assay/method) and one could add that margin to the context of variability/heterogeneity. However the link between to Q-profile of the biosimilar and the "comparability exercise" between biosimilar and reference biologic remains unclear! How does this relate to variability/heterogeneity and where is this incorporated into this guide? 3.2 Then I would like to highlight a general conceptual issue when taking the variation/heterogeneity and the life-resp. development-cycle of the reference biologic and the biosimilar into account. The main issue with the life- resp. development-cycle of the reference biologic and the biosimilar is, that safety and efficacy findings during the clinical comparability exercise could be controversial. This guide acknowledges already that the reference biologic might drift in its Q-profile, as part of its normal life-cycle. This bears the risk for the biosimilar developer that the QTPP might drift as well over the biosimilar development phase. For example for the clinical trial phase, it would be very advisable to conduct the comparability exercise against an unchanged (not drifted in its Q-profile) reference biologic, which served as QTPP for the biosimilar initially, if possible. As the time point if/when such a potential drift of the Q-profile of the reference biologic might occur cannot	

Stakeholder number *(To be completed by the Agency)*	**General comment (if any)**	**Outcome (if applicable)** *(To be completed by the Agency)*
	be foreseen by the biosimilar developer, the challenges and risks are considerable in the biosimilar developers eyes. (Always taken into account that the EMA already acknowledged the difficulties, that even minor differences in the Q-profile might have an impact on safety and efficacy, refer to below, for mAbs for example). *However, it may at the current stage of knowledge be difficult to interpret the relevance of minor quality differences in the physicochemical and biological characterization when comparing a biosimilar mAb to a reference mAb. EMA/CHMP/BMWP/403543/2010* I do not expect EMA to solve those case presented below, but would like to highlight potential complications and would like EMA to acknowledge those constrains. 3.2.1 Scenario 1: EMA permitted a process change for the reference biologic, which was considered acceptable. When compared head to head with the biosimilar in a clinical trial setting this change resulted in a measurable impact/difference (beyond the margins of equivalence). The Q-profile change of the reference biologic shifted, whereas the QTPP of the biosimilar was based on the pre-change Q-profile of the reference biologic. The post-change reference biologic had to be used for clinical trial comparability exercise due to normal shelf-life limitations.	

Stakeholder number *(To be completed by the Agency)*	General comment (if any)	Outcome (if applicable) *(To be completed by the Agency)*
	The biosimilar was shown to possess a better safety and efficacy profile compared to the reference biologic. (Example mentioned verbally in London during last meeting with EMA about: "Guideline on similar biological medicinal products containing monoclonal antibodies – non-clinical and clinical issues") • What is the consequence of that for the EMA? • What is the consequence of that for the reference product? • What is the consequence of that for the biosimilar? 3.2.2 Scenario 2: EMA permitted a process change for the reference biologic, which was considered acceptable. When compared head to head with the biosimilar in a clinical trial setting this change resulted in a measurable impact/difference (beyond the margins of equivalence). The Q-profile change of the reference biologic shifted, whereas the QTPP of the biosimilar was based on the pre-change Q-profile of the reference biologic. The post-change reference biologic had to be used for clinical trial comparability exercise due to normal shelf-life limitations. The reference was shown to possess a better safety and efficacy profile • Is the comparability exercise considered failed? • Is refining of the QTPP required and some re-development? 3.3 Conclusion: All stakeholders of the biosimilar	

Stakeholder number (To be completed by the Agency)	General comment (if any)	Outcome (if applicable) (To be completed by the Agency)
	application process will need to get familiar with the idea that their respective work and evaluations will be subject to unprecedented counterchecking.	

2. Specific comments on text

Line number(s) of the relevant text (e.g. Lines 20-23)	Stakeholder number (To be completed by the Agency)	Comment and rationale; proposed changes (If changes to the wording are suggested, they should be highlighted using 'track changes')	Outcome (To be completed by the Agency)
		Comment 1: **1st Comment:** As rightfully highlighted in this guide, the QTPP of the biosimilar is founded on the permitted variability of the reference biologic. This good direction given from this guide to the biosimilar developer to "understand the variability of the reference biologic first", is an important step to successfully mimic the quality profile (Q-profile) of the reference biologic by using the QTPP approach (at least in an initial phase of the biosimilar development program, refer to later comment). 1.1 The aims of this guide for the Q-profile of the biosimilar are aligned with the principle that similarity should be achieved but not necessarily a bio-better. This principle was repetitively highlighted by EMA, refer to below for example:	

Line number(s) of the relevant text *(e.g. Lines 20-23)*	Stakeholder number *(To be completed by the Agency)*	Comment and rationale; proposed changes *(If changes to the wording are suggested, they should be highlighted using 'track changes')*	Outcome *(To be completed by the Agency)*
		..."The biosimilarity exercise follows the main concept that clinical benefit has already been established by the reference medicinal product, and that the aim of a biosimilar development programme is to establish similarity to the reference product, not clinical benefit"... EMA/CHMP/BMWP/572643/2011 *A biosimilar mAb should be similar to the reference mAb in physicochemical and biological terms. Any observed relevant difference would have to be duly justified and could contradict the biosimilar principle. For quality aspects the principles as laid out in the guidelines on biosimilars including the "Guideline on similar biological medicinal products containing biotechnology-derived proteins as active substance: Quality issues" (EMEA/CHMP/BWP/49348/2005), and the "Guideline on development, production, characterisation and specifications for monoclonal antibodies and related substances" (EMEA/CHMP/BWP/157653/2007) apply. EMA/205886/2012* 1.2 If the reference biologic is "relative" heterogenic in its Q-profile which is likely to be the case with increasingly large and glycosilated biologics, the chances are considerable due to improvements in purification and analytical technologies that the biosimilar developer will have the following two choices: 1) to mimic the (within a lot)	

Line number(s) of the relevant text *(e.g. Lines 20-23)*	Stakeholder number *(To be completed by the Agency)*	Comment and rationale; proposed changes *(If changes to the wording are suggested, they should be highlighted using 'track changes')*	Outcome *(To be completed by the Agency)*
		variation of the reference medicinal drug substance 2) to focus on homogeneity/purity of the biosimilar drug substance and as a consequence to aim potentially for a Q-profile which is better than the one of the reference biologic.	
line 94		The approach 1) might be technically more challenging as more process knowledge is required, to mimic the biological diversity of the reference drug substance (but would be in line with my understanding of the requirements for Biosimilar), whereas the approach 2) bears the risk that the safety and efficacy profile will be different and therefore the Q-attributes not the same, compared to the reference medicinal product. 1.3 With regard to that, one sentence should be amended to make this point clearer, as otherwise the biosimilar developer has a development choice on which he should be given some guidance: ...*"performance and consistency of the manufacturing process of the biosimilar on its own"*... (Line 94)	
section 4 line 107 for example		Particular for biologics the process and the product are closely interlinked therefore this sentence can be well interpreted as: performance and consistency of the biosimilar itself, as the process output (product or drug substance). Consistency of the biosimilar then leads to think of homogeneity/purity and the reader is confused when compared to sections such as:	

Line number(s) of the relevant text *(e.g. Lines 20-23)*	Stakeholder number *(To be completed by the Agency)*	Comment and rationale; proposed changes *(If changes to the wording are suggested, they should be highlighted using 'track changes')*	Outcome *(To be completed by the Agency)*
		..."The applicant should demonstrate that the desired product and product-related substances present in the finished product of the biosimilar are highly similar to that of the reference medicinal product"...(Line 156, 157) **(I) Therefore a sentence at that stage (line 94) shall state that:** ***"the initial or cardinal biosimilar development aim is to mimic the quality profile of the reference biologic drug substance including its variation/heterogeneity (on a lot basis*1) where possible."*** 1*Taken into account that individual lots of reference biologic and biosimilar will be compared at clinical trial stage with the aim to show comparability. A purer/more consistent biosimilar could have different kinetics and clinical performance compared to the reference biologic, therefore the development should focus on mimicking the Q-profile of the reference biologic at an initial stage. 1.4 Further a sentence is required to clarify that through successive analysis and characterization (correlation of critical Q-attributes to physico-chemical-, biological-properties and the manufacturing process) a more consistent/purer (less heterogenic) Q-profile could be chosen for the biosimilar drug substance, compared to the reference biologic drug substance as long as the safety- and efficacy-profile are comparable (within the margin of equivalence). **(II)Therefore EMA should state that for in**	

Line number(s) of the relevant text *(e.g. Lines 20-23)*	Stakeholder number *(To be completed by the Agency)*	Comment and rationale; proposed changes *(If changes to the wording are suggested, they should be highlighted using 'track changes')*	Outcome *(To be completed by the Agency)*
		section 4 line 107 for example: ***"the better the correlation of molecular characteristic (including process understanding) of the biosimilar drug substance with quality attributes is understood, the more the biosimilar developer can shift the focus (of this development strategy) from mimicking the Q-profile of the reference biologic including its variability/heterogeneity; towards manufacturing a more consistent/pure biosimilar drug substance and therefore eliminating some of the heterogeneity of the reference biologic, as long the safety and efficacy remain comparable within the margins of equivalence"*** 1.5 By adding **(I)** and **(II)** to this guide, all ambiguity is removed. It gives a comprehensive solution for the development strategy of the biosimilar developer and aligns different sections within this guide, which per se read may be understood to be conflicting with each other. I believe that this reflects the intention of EMA. Proposed change (if any): please see above bold highlighted sections.	
		Comment: 2 ***2nd Comment:*** The scope of ICH Q8 R2 is the life-cycle of a drug, including the whole development	

Line number(s) of the relevant text *(e.g. Lines 20-23)*	**Stakeholder number** *(To be completed by the Agency)*	**Comment and rationale; proposed changes** *(If changes to the wording are suggested, they should be highlighted using 'track changes')*	**Outcome** *(To be completed by the Agency)*
		program. The QTPP (based on ICH Q8 R2) is a combination of mainly quality-attributes, but takes also into account safety- and efficacy-attributes. Therefore when utilizing the QTPP approach this is not limited to the quality section of the drug substance, as utilized for this guide, refer to Note 1 below: *Note 1* *Classically the QTPP consist of the following attributes with relation to:* *Quality:* • *Physical and chemical attributes of the dosage form,* • *cosmetic elements such as shape, color, size,* • *Container* • *etc.* *Safety:* • *Impurities and degradation products* • *etc.* *Efficacy:* • *Clinical performance,* • *Biopharmaceutical and pharmacokinetics* • *etc.* 2.1 It is clear that some of the Q-attributes can only be partially predicted and tested by the methodologies mentioned in this guide (due to the only predictive nature for those tests), refer to below.	

Line number(s) of the relevant text *(e.g. Lines 20-23)*	Stakeholder number *(To be completed by the Agency)*	Comment and rationale; proposed changes *(If changes to the wording are suggested, they should be highlighted using 'track changes')*	Outcome *(To be completed by the Agency)*
section 4		*Various assays have been established in the past years that allow for more in-depth characterisation of complex proteins, both on a physicochemical and a functional level, e.g. with potency assays, and there is experience in the assessment of minor quality differences due to changes in manufacturing processes for monoclonal antibodies. However, it may at the current stage of knowledge be difficult to interpret the relevance of minor quality differences in the physicochemical and biological characterization when comparing a biosimilar mAb to a reference mAb. EMA/CHMP/BMWP/403543/2010* EMA therefore acknowledges the need to confirm those predictions in clinical trials. As QTPP terminology was consciously chosen by EMA for this guide. This should be more incorporated in this guide by EMA, by making the link with guidance documents such as *"Guideline on similar biological medicinal products containing monoclonal antibodies – non-clinical and clinical issues" (EMA/CHMP/BMWP/403543/2010*). **(III) therefore EMA should state in section 4 that:** **"...at later development stages the established Q-attributes require confirmation in clinical trials and may be amended/completed, as the trial outcome cannot be foreseen. This could result in adjustment of the QTPP."** This statement would satisfy better the application and scope of ICH Q R2 for this guide.	

Line number(s) of the relevant text *(e.g. Lines 20-23)*	Stakeholder number *(To be completed by the Agency)*	Comment and rationale; proposed changes *(If changes to the wording are suggested, they should be highlighted using 'track changes')*	Outcome *(To be completed by the Agency)*
		Proposed change (if any): please see above bold highlighted sections.	

Appendix: 3

Questionnaire 1.1

Refer also to 3.3.1 Practical aspects of the questionnaires and critical considerations for related detail.

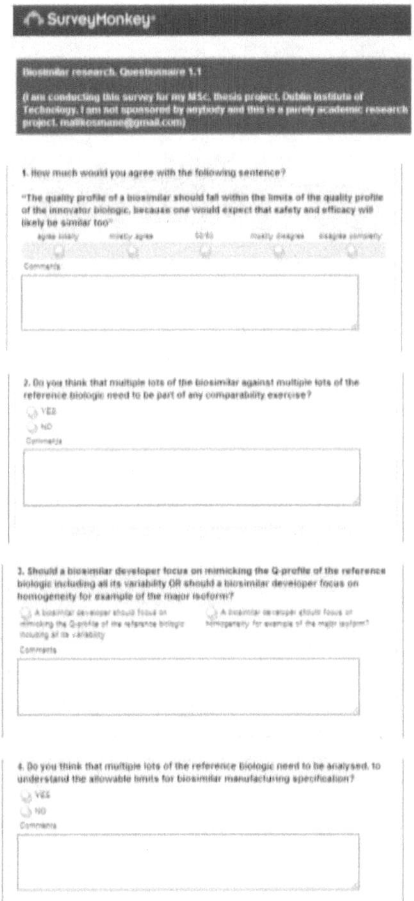

SurveyMonkey

Biosimilar research. Questionnaire 1.1

(I am conducting this survey for my MSc. thesis project, Dublin Institute of Technology. I am not sponsored by anybody and this is a purely academic research project. malikosmane@gmail.com)

1. How much would you agree with the following sentence?

"The quality profile of a biosimilar should fall within the limits of the quality profile of the innovator biologic, because one would expect that safety and efficacy will likely be similar too"

agree totally | mostly agree | 50/50 | mostly disagree | disagree completely

Comments

2. Do you think that multiple lots of the biosimilar against multiple lots of the reference biologic need to be part of any comparability exercise?

YES

NO

Comments

3. Should a biosimilar developer focus on mimicking the Q-profile of the reference biologic including all its variability OR should a biosimilar developer focus on homogeneity for example of the major isoform?

A biosimilar developer should focus on mimicking the Q-profile of the reference biologic including all its variability

A biosimilar developer should focus on homogeneity for example of the major isoform?

Comments

4. Do you think that multiple lots of the reference biologic need to be analysed, to understand the allowable limits for biosimilar manufacturing specification?

YES

NO

Comments

*** 5. Would you agree with the following:**

"the initial or cardinal biosimilar development aim is to mimic the quality profile of the reference biologic drug substance including its variation/heterogeneity (on a lot basis) where possible."

"the better the correlation of molecular characteristic (including process understanding) of the biosimilar drug substance with quality attributes is understood, the more the biosimilar developer can shift the focus (of this development strategy) from mimicking the Q-profile of the reference biologic including its variability/heterogeneity, towards manufacturing a more consistent/pure biosimilar drug substance and therefore eliminating some of the heterogeneity of the reference biologic, as long the safety and efficacy remain comparable within the margins of equivalence"

- ◯ YES
- ◯ NO

Comments

6. Would you recommend to the biosimilar developer to initiate clinical trials (i.e. clinical comparability exercise), if the Q-profile of the reference biologic shifted and the QTPP of the biosimilar is based on the "old Q-profile" of the reference biologic

- [ ] Yes, as the risk of detecting a clinically significant difference due to changes of the Q-profile is relatively low risk for the biosimilar developer
- [ ] Yes, as the "new Q-profile" of the reference biologic was approved by EMA and therefore should fall within similar safety and efficacy margins as the "old Q-profile". Thus, it should not impact of the clinical comparability exercise
- [ ] Yes, as in the case where differences in clinical comparability exercise were detected Q-profile of the biosimilar can be adjusted rapidly to follow the most recent Q-profile of the reference biologic
- [ ] No, as the Q-profile of the biosimilar should have prior the clinical comparability exercise a very similar Q-profile compared to the reference biologic. This should result in a similar clinical outcome
- [ ] No, as differences during the comparability beyond the margin of tolerance would require additional investigation and the burden of those is deemed as not desirable
- [ ] No, as the risk of detecting a clinically significant difference due to changes of the Q-profile is relatively high risk for the biosimilar developer

Comments

7. Do you test multiple lots of the reference biologic against multiple lots of the biosimilar during the clinical efficacy exercise?

- ◯ YES
- ◯ NO

Comments

8. Even if the mechanism of action is identical, could the efficacy be different as the isoform ratio may be different between biosimilar and reference biologic?

- ◯ YES
- ◯ NO

Comment

*** 9. Please indicate your academic and professional background**

	Academic background 1	Academic background 2	Professional background 1	Professional background 2
Academic and professional background	▼	▼	▼	▼

Q10

10. Facultative contact details

Name

e-mail

Appendix: 4

Questionnaire 1.2

Refer also to 3.3.1 Practical aspects of the questionnaires and critical considerations for related detail.

The benefit of participating in the questionnaire is that you can get sections of my study on biosimilars and their regulatory caveats related to those follow on biologics. (**If you participate and fill in all essential sections, I will send you via e-mail exclusively those findings).**
Please forward me the filled-in questionnaire via e-mail to me

Name (not essential):	**Title/Role** (example QC manager, Prof):
Organisation (not essential):	**E-mail address:**

Academic background 1: Technician ☐ Degree ☐ Masters ☐ PhD ☐ Professor ☐
Academic background 2: Bio. ☐ Chem. ☐ Pharma. ☐ Medicine ☐ Biotech. ☐ Technical ☐ Other ☐
Professional background 1: Academia ☐ Goverment Agency ☐ Pharmaceutical ☐ Other ☐
Professional background 2: Technical ☐ Science & Laboratory ☐ Q.C. ☐ Q.A. ☐ R.A. ☐ Other ☐

Professionally intersted in: Innovator biologics (i.e. Original biologic drugs) ☐ Biosimilars (follow on biologics) ☐ Both ☐ Don't know what a biosimilar is ☐ Not interested in biologics at all ☐

Note on comment section: **Comments are highly welcomed and appreciated. Any idea, comment, reference or opinion is very welcome! Please try to give at least one comment per question.**
Terminology: For simplification purposes the **quality profile (Q-profile)** in this questionnaire is defined as the full characterisation of a biologic. (The quality profile results from thorough analyses and is a measure of the process output and therefore a definition of the variability of multiple lots of the biologic drugs)
Statement: This questionnaire is for academic purposes only. There is no commercial interest behind this questionnaire. Thus, your details won't be made accessible with exception of academic staff (D.I.T.).

1.0 General questions:

1.1) Do you think that within one lot of a complex biologic (such as a mAb), it will always contain a variations of mAb molecules (due to product or process related heterogeneity), which are slightly different to each other?

Yes / No	Comments:

1.2) Do you think that the ratio of those slightly different (mAb)-molecules is unique for an individual lot.

Yes / No	Comments:

1.3) Do you think that between multiple lots of a complex biologic (such as mAb), there will be always lot to lot variation of the Q-profile (due to product or process related variation and/or process changes)?

Yes / No	Comments:

1.4) What are the major the 5 major factors contributing to variation and heterogeneity of a biologic (such as mAb) in your opinion?

Rated by significance

most significant

1)____________________
2)____________________
3)____________________
4)____________________
5)____________________

less significant

Comments:

2.0 Detailed biosimilar questions:

2.1) How much would you agree with the following sentence?

"The quality profile of a biosimilar should fall within the limits of the quality profile of the innovator biologic, because one would expect that safety and efficacy will likely be similar too"

Rate the level of agreement: 1 2 3 4 5 (please encircle your rating)

agree totally ⟷ disagree completely

Comments:

2.2) Do you think that multiple lots of the reference biologic need to be analysed, to understand the allowable limits for biosimilar manufacturing specification?

Yes / No Comments:

2.3) Should a biosimilar developer focus on mimicking the Q-profile of the reference biologic including all its variability **or** focus on homogeneity for example of the major isoform?

Please elaborate your thoughts and comment!

2. 4) Do you think that multiple lots of the biosimilar against multiple lots of the reference biologic need to be part of any comparability exercise?

Yes / No	Please elaborate your thoughts and comment!

Comments:

2.5) Would you recommend to the biosimilar developer to initiate clinical trials (i.e. clinical comparability exercise), if the Q-profile of the reference biologic shifted and the QTPP of the biosimilar is based on the "old Q-profile" of the reference biologic.

- ☐ Yes, as the risk of detecting a clinically significant difference due to changes of the Q-profile is relatively low risk for the biosimilar developer.
- ☐ Yes, as the "new Q-profile" of the reference biologic was approved by EMA and therefore should fall within similar safety and efficacy margins, as the "old Q-profile". Thus it should not impact of the clinical comparability exercise.
- ☐ Yes, as in the case where differences in clinical comparability exercise were detected Q-profile of the biosimilar can be adjusted rapidly to follow the most recent Q-profile of the reference biologic.
- ☐ No, as the Q-profile of the biosimilar should have prior the clinical comparability exercise a very similar Q-profile compared to the reference biologic. This should result in a similar clinical outcome.
- ☐ No, as differences during the comparability beyond the margin or tolerance would require additional investigation and the burden of those is deemed as not desirable.
- ☐ No, as the risk of detecting a clinically significant difference due to changes of the Q-profile is relatively high risk for the biosimilar developer.

Comments:

2.6) What do you think that about meta analysis within the biosimilar context? Please indicate where you agree and comment.

- ☐ Meta analysis should not be required, as the clinical comparability exercise should be clearly designed with a clear outcome.
- ☐ Meta analysis should not be required, as the summary of data generated for the comparability exercise are not usually not sufficiently large to be used in a statistical meta analytical way.
- ☐ Mentioning of Meta analysis by EMA may indicates that EMA foresees situations where the clinical comparability exercise is not clear based on head to head comparisons.
- ☐ Mentioning of Meta analysis by EMA may indicate that EMA foresees situations where the summary of clinical data are substancial enough to conduct Meta analysis.
- ☐ Meta analysis is to welcome, as it allows to combine data and to re-assess the outcome. Thus more confidence in the clinical comparability exercise is generated.
- ☐ Meta analysis is to welcome, as it is also be part of many innovator biologic clinical trial data analysis. Therefore Meta analysis within the biosimilar context appear to be normal.

Comments:

2.7) Do you think that for biosimilar comparability exercise the classical phase I, phase II and phase III approach is required? Please give your rational for or against that.

Yes / No	Please elaborate your thoughts and comment!

Comments:

EMA recommends clinical comparability studies for mAbs "Guideline on similar biological medicinal products containing monoclonal antibodies –non clinical and clinical issues" (EMA/CHMP/BMWP/403543/2010).

2.8.1) Would it make sense to test multiple lots of the reference biologic against multiple lots of the biosimilar during the clinical PK/PD studies?

Yes / No	Please elaborate your thoughts and comment!

Comments:

2.8.2) Please comment if you would combine PK, PD studies (step 1) with clinical efficacy studies (step 2)

Yes / No	Please elaborate your thoughts and comment!

Comments:

2.8.3) Do you think that almost always the steps 2 (clinical efficacy) of the comparability exercise is required?

Yes / No	Please elaborate your thoughts and comment!

Comments:

2.8.5) Do you need to conduct this clinical efficacy exercise as a head to head comparison?

Yes / No	Please elaborate your thoughts and comment!

Comments:

2.8.6) Do you test multiple lots of the reference biologic against multiple lots of the biosimilar during the clinical efficacy exercise?

Yes / No	Please elaborate your thoughts and comment!

Comments:

2.9) Extrapolation of indications. EMA suggests extrapolating indications if the mechanism of action is identical. Do you know of any biosimilar cases where this happened? Please cite, if possible.	
2.10) Even if the mechanism of action is identical, could the efficacy be different as the isoform ratio may be different between biosimilar and reference biologic?	
Yes / No	Please elaborate your thoughts and comment!
Comments:	

Appendix: 5

Answers from the EMA to the "Questions and comments submitted to the EMA for the "Guideline on similar biological medicinal products containing monoclonal antibodies Draft".

References	**Question or comment submitted** "Guideline on similar biological medicinal products containing monoclonal antibodies Draft"	**Notes and clarification, if required**	**Answer from EMA** Overview of comments received on 'Guideline on similar biological medicinal products containing monoclonal antibodies' (EMA/CHMP/BMWP/403543/2010)
	The EMA guide EMEA/CHMP/BWP/49348/2005 and subdivides variability in: i) Product variability ii) Process related variability. Those two variability types form what is called the "Quality Profile" of the Biosimilar. However these types of variability are relevant for all biotechnological derived medicinal products, be it the Reference Medicinal Product or be it the Biosimilar. One particular issue is in my opinion the variability of the Reference Medicinal Product itself. Refer to Note: 1 1.) Therefore the variability of the Reference Medicinal Product itself (as a variable) leads to questions on how to fulfil requirements for the Biosimilar i.e. to demonstrate comparable	Note: 1 This "overlaying issue of product heterogeneity and variation" is founded on and also (partially) reflected in the following documents: *1.) "Concept paper on the revision of the guideline on similar biological medicinal products containing biotechnology-derived proteins as active substance: quality issues".* *Section 3 (Problem statement)* Which takes into account the life cycle of the Biosimilar and variations of the process, impacting the variability	*The "quality profile" is these two sorts of variability PLUS the usual stand-alone physicochemical and biological characterisation PLUS a comprehensive comparability exercise. As such, the "quality programme" for a biosimilar is more extensive than that of a stand-alone.* *The "overarching" guideline and the general nonclinical/clinical guideline are subject to revision.* EMA/205886/2012, Page 30/422, Stakeholder 13

<table>
<tr>
<td></td>
<td>clinical efficacy and safety versus the Reference Medicinal Product.

Or differently expressed:

2.) When requiring any direct comparison, be it in a non-clinical or clinical setting, batch to batch variation of the Reference Medicinal Product (due to the glycosylation heterogeneity of mAbs or process impurities associated with the production of mAbs i.e. the full quality profile of the Reference Medicinal Product) and the quality profile of the Biosimilar need to be taken into account when designing studies and analyzing data.

I call this train of thoughts in this submission the "overlaying issue of product heterogeneity and variation (1)".</td>
<td>(Quality Profile) of the Biosimilar. This is of relevance for any Biological Medicinal Product, be it a Reference Medicinal Product or Biosimilar.

This "overlaying issue of product heterogeneity and variation [(1)]", appears also to be justified taken into account the exiting guidance:

2.) Doc. Ref. EMEA/CHMP/BMWP/101695/2006
GUIDELINE ON COMPARABILITY OF BIOTECHNOLOGY-DERIVED MEDICINAL PRODUCTS AFTER A CHANGE IN THE MANUFACTURING PROCESS NON-CLINICAL AND CLINICAL ISSUES
"Demonstration of comparability is a sequential process, beginning with quality studies (limited or comprehensive) and supported, as necessary, by non-clinical and/or clinical studies.
This guideline will address the requirements for non-clinical and/or clinical bridging studies to demonstrate that the modification has no impact on safety and efficacy.
The selection of non-clinical versus clinical testing is product-driven, i.e. a model should be chosen that best detect clinically relevant differences with sufficient accuracy"

3.) Doc Ref: EMEA/CHMP/BWP/49348/2005
GUIDELINE ON SIMILAR BIOLOGICAL MEDICINAL PRODUCTS</td>
<td></td>
</tr>
</table>

		CONTAINING BIOTECHNOLOGY-DERIVED PROTEINS AS ACTIVE SUBSTANCE: QUALITY ISSUES *"Consequently, an extensive comparability exercise will be required to demonstrate that the biosimilar and reference products have similar profiles in terms of quality, safety and efficacy.* *Based on the comparability approach and when supported by sufficiently sensitive analytical systems, the demonstration of comparability at the quality level may connect the biosimilar product to the nonclinical and clinical data previously generated with the reference product"* *"It is recognised that each medicinal product is defined by the molecular composition of the active substance resulting from its process, which may introduce its own process related impurities.* *Consequently, the biosimilar product is defined by the following two sets of characteristics: i) related to the characteristics of the molecule (including product related substances/impurities), and ii) related to its process"* *4.) CHMP/437/04* *GUIDELINE ON SIMILAR BIOLOGICAL MEDICINAL PRODUCTS* *"Biological medicinal products are usually more difficult to characterise than chemically derived medicinal products. In addition, there is a spectrum*	

		of molecular complexity among the various products (recombinant DNA, blood or plasma-derived, immunologicals, gene and cell-therapy, etc.). *Moreover, parameters such as the three-dimensional structure, the amount of acido-basic variants or post-translational modifications such as the glycosylation profile can be significantly altered by changes, which may initially be considered to be 'minor' in the manufacturing process. Thus, the safety/efficacy profile of these products is highly dependent on the robustness and the monitoring of quality aspects.* *Therefore:* *– The standard generic approach (demonstration of bioequivalence with a reference medicinal product by appropriate bioavailability studies) is normally applied to chemically derived medicinal products. Due to the complexity of biological/biotechnology-derived products the generic approach is scientifically not appropriate for these products. The " similar biological medicinal products" approach, based on a comparability exercise, will then have to be followed"*	
Referring to lines: 135 to 138	*130 4.1. In vitro pharmacodynamic (PD) studies = step1* *131 In order to assess any difference in biological activity between the similar biological medicinal and the*		

	132 reference medicinal product, data from a number of comparative in vitro studies, some of which may *133 already be available from quality-related assays, should be provided.* Comment: Depending on assay sensitivity and methodology, comparative in vitro studies will always reveal differences in parameters cited in lines 135 to 138 of the draft guide. This is based on the reasonable assumption (based on available publications), that even the Reference Medicinal Product itself will vary between every batch, in for example the ratio glycosylated proteins and the level of glycosylation, refer also to [(1)]. Therefore the cited parameters will vary (slightly) for each batch of the Reference Medicinal Product or the Biosimilar, (again assuming sufficiently high assay sensitivity and methodology). Proposed change (if any): A statement at the beginning of section 4.0 should incorporate the following: 1.) "the Biosimilar requires to such an extend similarity to the Reference Medicinal Product, as the Reference Medicinal Product is similar to itself, when compared	 By adopting this proposed approach, the level of comparability between the Reference Medicinal Product and the Biosimilar, using sound, measurable and justified parameters will reduce risks associated with clinical trials and will likely lead to (at least) a similar efficacy and safety profile of the Biosimilar in clinical trials compared to the Reference	*Proposed changes not accepted.* *These topics are addressed in the " Guideline on similar biological medicinal products containing biotechnology-derived proteins as active substance: quality issues " which is currently under revision.* *EMA/205886/2012, Page 154/422, stakeholder 13*

	on a batch to batch basis, as the optimum. Deviation of this approach is generally acceptable, but requires justification and possibly additional clinical evaluation in order to assess associated risks" **Further to guide the industry and to avoid unnecessary costs and disappointments, it should be stated that:** **2.)** "For non clinical setting it is advisable that the "Quality Profile", (refer to [1]) of the Biosimilar is within the levels of variation observed for the "Quality Profile" of the Reference Medicinal Product. Thoroughly determination of the "Quality Profile" of the Reference Medicinal Product can be used to set the margins of tolerance (for variation) for the Biosimilar "Quality Profile". Deviation of this approach is generally acceptable, but requires justification and possibly additional clinical evaluation in order to assess associated risks".	Medicinal Product	
References	**Question or comment submitted** **"Guideline on similar biological medicinal products containing monoclonal antibodies Draft"**	**Notes and clarification, if required**	**Answer from EMA** **Overview of comments received on 'Guideline on similar biological medicinal products containing monoclonal antibodies' (EMA/CHMP/BMWP/403543/2010)**
Referring to lines: 209 to 212	*209 The primary objective of the pharmacokinetic studies performed to support a Marketing Authorisation*		*Not accepted. This is not specific to mAbs and will be addressed in the revision of the general Non-clinical and Clinical guideline.*

Referring to lines: 239 and 240 Referring to lines: 254 to 256	*210 Application (MAA) for a similar biological medicinal product is to show comparability in* *211 pharmacokinetics of the biosimilar with the reference product in a sufficiently sensitive and* *212 homogeneous population.* *239 clearance and half-life should be determined and reported in a descriptive manner. If relevant* *240 differences occur the assumption of similar PK might be seriously questioned. If such results are* *As a principle, any* *254 widening of the conventional equivalence margin beyond 80-125% requires thorough justification,* *255 including an estimation of potential impact on clinical efficacy and safety. This should be discussed with* *256 regulatory authorities.* Comment: For the clinical aspects the same point raised in [Note:1] is valid. 1.) Assuming the Biosimilar needs to be as safe and effective as the Reference Medicinal Product as indicated in lines: 209 to 212 and that the clinical efficacy of the Biosimilar is required to be similar to the Reference Medicinal Product.	Note: 1 "overlaying issue of product heterogeneity and variation"	*EMA/205886/2012, Page 229/422, stakeholder 13*

	2.) The PK studies require to be similar compared to the Reference Medicinal Product. Referring to lines: 239 and 240 3.) Taken also lines: 254 to 256 into account, clearly the safety and efficacy comparability exercise is subject to tolerance margins. However this margin of tolerance is not clear and would require further elucidation. What is the justification for a "generally acceptable" margin with regard to Biosimilars? Taken into consideration the following points: **1.) the problems associated with populations for trials for example in a cancer setting where sufficiently homogenous population might be rare and healthy volunteers cannot be used for ethical reasons either.** **2.) cases where the "Quality Profile" of the Reference Medicinal Product exceeds already an efficacy margin of 80-125% when compared on a batch to batch basis in clinical trial to itself.** **There is also additional clarification required, from the innovator industries point of view, mainly on the following points:** 1.) Taken the costs associated with clinical trials into account and the need to reduce healthcare		

	costs, is it required to assess multiple batches of the Reference Medicinal Product when comparing it with the Biosimilar in a clinical trial setting, as a general question? 2.) As indicated in lines: 209 to 212 the expectation is that safety and efficacy are "comparable". A) Does this mean the acceptable tolerance margin for the Biosimilar in a clinical efficacy trial is required to be within the variation of endpoints of multiple batches of the Reference Medicinal Product and then the conventional equivalence margin of 80%-125% for the Biosimilar is applicable? Or B) Does this mean the acceptable tolerance margin for the Biosimilar in a clinical efficacy trial is based on one batch the Reference Medicinal Product and then the conventional equivalence margin of 80-125% for the Biosimilar is applicable?		
References	**Question or comment submitted** **"Guideline on similar biological medicinal products containing monoclonal antibodies Draft"**	**Notes and clarification, if required**	**Answer from EMA** Overview of comments received on 'Guideline on similar biological medicinal products containing monoclonal antibodies' (EMA/CHMP/BMWP/403543/2010)
	Extrapolation of indication for a Biosimilar The explicit mentioning of and encouraged usage of "Meta Analysis" in this		*However, the BMWP feels that terms like "conventional equivalence margin" do not require further explanation, since they are*

	guide However as in any progressive document, there are few areas which require further clarification. Although these areas are not specifically linked to mAbs and moreover to Biosimilars in a more general way, this guide uses some key terminology and concepts such as: 1. Extrapolation of Indications 2. Meta Analysis 3. Conventional Equivalence Margin , which require further explanation. I feel it would be appropriate either to clarify those terms and concepts as part of this guide or to address them in a separate guide, applicable for all Biosimilars taking into account the heterogeneity and the nature of mAbs (and other Biosimilars).		*either defined in other CHMP guidance being referred to by this guidance, or where an explanation is given (e.g. in the mAbs guidance: "widening of the conventional equivalence margin beyond 80-125% requires thorough justification", where 80-125% is the definition of "conventional").* EMA/205886/2012, Page 30/422, Stakeholder 13
Referring to lines: 424 to 428	*424 6. Extrapolation of Indications* *425 Extrapolation of clinical efficacy and safety data to other indications of the reference mAb, not* *426 specifically studied during the clinical development of the biosimilar mAb, is possible based on the* *427 overall evidence of biosimilarity provided from the comparability exercise and with adequate* *428 justification.*	The EMA wrote on this topic the following: 1.) *2 EMA/CHMP/BMWP/86289/2010* *3 Committee for Medicinal Products for Human Use (CHMP)* *4 Guideline on immunogenicity assessment of monoclonal* *5 antibodies intended for in vivo clinical use.* *6 Draft* *"261 MAbs exert their action*	Not commented upon by EMA, in EMA/205886/2012, nor is this submitted comment included or mentioned in EMA/205886/2012. Assumingly this comment was overseen by the EMA, as the author of this study did not mark the section sufficiently.

	Comment: Referring to lines 424 to 428, what would the justification be? As the mechanism of action of the Reference Medicinal Product could be different from indication to indication, is it justified to allow "Extrapolation of Indications" without showing the efficacy for this particular indication for the Biosimilar? For example for some indication an mAb might bind to a cell surface, whereas in other indications the primary action is to bind to soluble antigen. The subsequent clearance and inactivation mechanism might be very different. Differences in the glycosylation pattern for example between the Reference Medicinal Product and Biosimilar could be for one mechanism of action negligible, whereas for another mechanism of action very important. The mechanism of action can be different from indication to indication. As the mechanism of action can be influenced by the glycosylation pattern, which can be influenced by the "Quality Profile" (refer to "overlaying issue of product heterogeneity and variation [(1)]"); an important consequence of this train of thoughts is that new clinical efficacy data would be required for a new	*by various mechanisms ranging from simple binding to antigen, which alone 262 mediates the clinical effect, to binding antigen and mediating one or more immunobiological 263 mechanisms which combine to produce the overall clinical response. Therefore, although simple 264 binding may seem to be the only mechanism operating to achieve clinical efficacy, other effects may 265 also play a role in this."* *And* *"the mode of action of the mAb (e.g., lytic, apoptotic), and especially the nature of 331 the target molecule (e.g., immunosilencing, immunostimulating), needs to be adequately characterized 332 and comprehensively investigated"*	

	indication in order to ascertain that efficacy of the Biosimilar is comparable to the efficacy of the Reference Medicinal Product for this new indication (assuming the case of a different mechanism of action).		

www.ingramcontent.com/pod-product-compliance
Lightning Source LLC
LaVergne TN
LVHW102226300926
849160LV00005B/45
9783954891870